AF248742

THE SEARCH
FOR OIL

STATISTICS

Textbooks and Monographs

A SERIES EDITED BY

D. B. OWEN, *Coordinating Editor*

Department of Statistics
Southern Methodist University
Dallas, Texas

PAUL D. MINTON

Virginia Commonwealth University
Richmond, Virginia

JOHN W. PRATT

Harvard University
Boston, Massachusetts

VOLUME 1: The Generalized Jackknife Statistic
H. L. Gray and W. R. Schucany

VOLUME 2: Multivariate Analysis
Anant M. Kshirsagar

VOLUME 3: Statistics and Society
Walter T. Federer

VOLUME 4: Multivariate Analysis:
A Selected and Abstracted Bibliography, 1957-1972
Kocherlakota Subrahmaniam and Kathleen Subrahmaniam

VOLUME 5: Design of Experiments: A Realistic Approach
Virgil L. Anderson and Robert A. McLean

VOLUME 6: Statistical and Mathematical Aspects of Pollution Problems
edited by John W. Pratt

VOLUME 7: Introduction to Probability and Statistics (in two parts).
Part I: Probability. Part II: Statistics
Narayan C. Giri

VOLUME 8: Statistical Theory of the Analysis of Experimental Designs
J. Ogawa

VOLUME 9: Statistical Techniques in Simulation (in two parts)
Jack P. C. Kleijnen

VOLUME 10: Data Quality Control and Editing
Joseph I. Naus

VOLUME 11: Cost of Living Index Numbers: Practice, Precision,
and Theory
Kali S. Banerjee

VOLUME 12: Weighing Designs. For Chemistry, Medicine, Economics,
Operations Research, Statistics
Kali S. Banerjee

VOLUME 13: The Search for Oil: Some Statistical Methods
and Techniques
edited by D. B. Owen

OTHER VOLUMES IN PREPARATION

THE SEARCH FOR OIL

SOME STATISTICAL METHODS AND TECHNIQUES

edited by

D. B. OWEN

Department of Statistics
Southern Methodist University
Dallas, Texas

MARCEL DEKKER, INC.   New York

MARCEL DEKKER, INC.
270 Madison Avenue, New York, New York 10016

LIBRARY OF CONGRESS CATALOG CARD NUMBER: 75-25162
ISBN: 0-8247-6342-4
Current printing (last digit):
10 9 8 7 6 5 4 3 2 1

PRINTED IN THE UNITED STATES OF AMERICA

CONTRIBUTORS TO THIS VOLUME

Kamal C. Chanda, Department of Mathematics, Texas Tech University,
Lubbock, Texas

Herbert T. Davis, Department of Mathematics and Statistics,
University of New Mexico, Albuquerque, New Mexico

Manus R. Foster, Mobil Research and Development Corporation,
Dallas, Texas

Henry L. Gray, Southern Methodist University, Dallas, Texas

L. F. Guseman, Jr., Department of Mathematics, Texas A&M University, College Station, Texas

L. H. Koopmans, Department of Mathematics and Statistics,
University of New Mexico, Albuquerque, New Mexico

Anant M. Kshirsagar, Institute of Statistics, Texas A&M University, College Station, Texas

Jack Chao-sheng Lee, Department of Mathematics, Wright State
University, Dayton, Ohio

Emanuel Parzen, Statistical Science Division, State University
of New York, Buffalo, New York

Clifford Qualls, Department of Mathematics and Statistics,
University of New Mexico, Albuquerque, New Mexico

G. P. Steck, Numerical Mathematics Division, Sandia Laboratories,
Albuquerque, New Mexico*

John W. Van Ness, Institute for Mathematical Sciences, University
of Texas, Dallas, Texas

Homer F. Walker, Department of Mathematics, University of Denver,
Denver, Colorado†

*Presently with the Reactor Safety Studies Division
†Present address: Department of Mathematics, University of
Houston, Houston, Texas

PREFACE

In the fall of 1973 two symposia were held at Southern
Methodist University. The first symposium concentrated on time
series modeling and analysis. The second symposium revolved
around classification problems. This volume is a compilation
of the papers presented at the two symposia.

A time series problem may be defined as one in which data
are gathered in time sequence. A good deal of oil exploration
data is taken in time sequence and is amenable to analysis by
the techniques discussed here.

A solution to a classification problem may be defined as a
process of assigning a new object to one of several possible
populations. For instance, based on geological data, seismic
data and other observable data we may want to classify a new
well site as a possible oil producer, a possible gas producer,
or as neither.

Of course, the big problem here, and the one for which oil
companies guard their solutions most carefully, is which variables are the best predictors of success. These company secrets
are not discussed here. However, the use of the techniques discussed here give the best solutions known at this time to the
problems once the proper variables to measure have been identified.

John Burg described his technique for analyzing time series
data during the first symposium. However, since he was at that
time pursuing a Ph.D. degree at Stanford University, he declined
to contribute to the symposium proceedings. Dr. Manus Foster
and Dr. Henry L. Gray studied Burg's work and agreed to fill the
gap created by the absence of any paper by Burg. The final
paper in these proceedings is a summary and assessment of Burg's

1975 Ph.D. dissertation accepted by Stanford. In addition, Gray and Foster compare the method of Burg to similar results in the literature and suggest a new spectral estimator for consideration.

The funds for bringing the symposia speakers to Dallas were derived from a grant from the Mobil Foundation, Inc. The Department of Statistics wishes to express its gratitude to the Mobil Foundation for its generosity.

D. B. Owen

July 1975

CONTENTS

Contributors to This Volume   iii

Preface   v

*SOME SOLUTIONS TO THE TIME SERIES MODELING AND
PREDICTION PROBLEM* Emanuel Parzen   1

*CROSSING PROBABILITY BOUNDS FOR NONSTATIONARY
GAUSSIAN PROCESSES* L. H. Koopmans Clifford Qualls   17

MULTIVARIATE PREDICTION AND SPLINE FUNCTIONS
Herbert T. Davis   33

PROBLEMS IN SELECTING A PATTERN RECOGNITION ALGORITHM
John W. Van Ness   41

*ON MINIMIZING THE PROBABILITY OF MISCLASSIFICATION FOR
LINEAR FEATURE SELECTION: A CLASSIFICATION
PROCEDURE* L. F. Guseman, Jr. Homer F. Walker   61

A CLASS OF NONPARAMETRIC CLASSIFICATION RULES
Kamal C. Chanda Jack Chao-sheng Lee   83

SOME THOUGHTS ON PATTERN CLASSIFICATION BY ATTRIBUTES
G. P. Steck   121

THE ROLE OF CANONICAL VARIABLES IN DISCRIMINANT ANALYSIS
Anant M. Kshirsagar   147

AUTOREGRESSIVE, MAXIMUM ENTROPY, AND G-SPECTRAL ESTIMATION
Henry L. Gray Manus R. Foster   169

Index   191

THE SEARCH
FOR OIL

SOME SOLUTIONS TO THE TIME SERIES
MODELING AND PREDICTION PROBLEM

Emanuel Parzen

State University of New York at Buffalo
Buffalo, New York

1. INTRODUCTION

The art of time series modeling can be regarded as composed of two types of techniques which I call *hard* and *soft*.
If you were to ask me casually how to analyze a set of data of
interest to you, I would have to reply that it would depend on
the list of alternative models that should be considered, and
I could only determine that by looking at the data. A graduate
student upon hearing this will often comment, "That's the
trouble with time series analysis; it's a soft subject. What I
want is a hard subject where you can always give clear and un-
ambiguous advice (hard rules) on how to analyze the data."
The dilemma of my career has been my sympathy for this attitude.
On the one hand, I am committed to the value of hard rules
(that is, theorems) as the best basis for developing sound
soft rules. On the other hand, I have experienced that at the
firing line of working on real problems, one must wisely and
empirically develop soft rules, not completely justified by
theory, but justified by the fact they seem to work best or
well.

Real time series are usually nonstationary, and there are
not as yet hard rules for when to model a time series in terms
of time-varying means (regression against deterministic

functions of time, or decomposition into trend and seasonal) as
against correlations (autoregression). Even the autoregression
approach is not yet considered completely routine because of
the problem of determining the order of the autoregression.
In this paper I present an approach to solving the latter
problem.

We suppose that we have a finite sample $Y(t)$, t = 1, 2,
..., T, of a zero mean, stationary, normal time series
$\{Y(t), t = 0, \pm 1, \ldots \}$ with covariance function

$$R(v) = E[Y(t)Y(t + v)], \quad v = 0, \pm 1, \ldots \tag{1.1}$$

and spectral density function

$$f(\omega) = \frac{1}{2\pi} \sum_{v=-\infty}^{\infty} e^{-iv\omega} R(v) \tag{1.2}$$

2. PREDICTION AS A TIME SERIES DECOMPOSITION AND MODEL

To model a time series Y we write it

$$Y = Y^{\mu} + Y^{\nu} \quad \text{or} \quad Y(t) = Y^{\mu}(t) + Y^{\nu}(t) \tag{2.1}$$

where Y^{μ} denotes the part of Y explained by the explanatory
variables used, and Y^{ν} is the new (therefore, ν) part of
Y not explainable. For time series the notion of explain-
ability is made precise by the notion of *minimum mean square
error linear predictability*. When Y is normal, define

$$Y^{\mu}(t) = E[Y(t) \mid Y(t - 1), Y(t - 2), \ldots] \tag{2.2}$$

to be the conditional expectation (and, therefore, best predictor) of the value of $Y(t)$ given past values $Y(t - 1)$,
$Y(t - 2), \ldots,$ and

$$Y^{\nu}(t) = Y(t) - Y^{\mu}(t) \tag{2.3}$$

to be the error of (one-step ahead) prediction (we call $Y^{\nu}(t)$
the *innovation* at time t).

Formulas for $Y^{\mu}(t)$ can be expressed in many ways; a natural way is to write it as an infinite series of past values of the time series

$$-Y^{\mu}(t) = \sum_{j=1}^{\infty} \alpha_{\infty}(j) Y(t - j) \tag{2.4}$$

The minus sign on $-Y^{\mu}(t)$ is chosen to provide a convenient formula for $Y^{\nu}(t)$

$$Y^{\nu}(t) = \sum_{j=0}^{\infty} \alpha_{\infty}(j) Y(t - j) \tag{2.5}$$

defining $\alpha_{\infty}(0) = 1$.

The predictor coefficients $\alpha_{\infty}(j)$ are determined by the orthogonality conditions

$$E[Y^{\nu}(t) Y(t - k)] = 0, \quad k = 1, 2, \ldots \tag{2.6}$$

which yield an infinite set of linear equations, involving the covariances $R(j - k) = E[Y(t - j) Y(t - k)]$,

$$\sum_{j=0}^{\infty} \alpha_{\infty}(j) R(j - k) = 0, \quad k = 1, 2, 3, \ldots \tag{2.7}$$

called the Yule-Walker equations. The minimum mean square prediction error is given by

$$\sigma_{\infty}^{2} = E[|Y^{\nu}(t)|^{2}] = \sum_{j=0}^{\infty} \alpha_{\infty}(j) R(j) \tag{2.8}$$

Finite memory prediction plays a central role in time series modeling. For each integer $m = 1, 2, \ldots,$ one can define coefficients $\alpha_{m}(0) = 1, \alpha_{m}(1), \ldots, \alpha_{m}(m)$ satisfying the truncated system of Yule-Walker equations

$$\sum_{j=0}^{m} \alpha_{m}(j) R(j - k) = 0, \quad k = 1, 2, \ldots, m \tag{2.9}$$

These are the coefficients of the minimum mean square error
m-memory predictor

$$Y^{\mu,m}(t) = -\sum_{j=1}^{m} \alpha_m(j)Y(t - j) = E[Y(t)|Y(t - 1), \ldots, Y(t - m)].$$

$$(2.10)$$

The memory m minimum mean square prediction error is given by

$$\sigma_m^2 = E[|Y(t) - Y^{\mu,m}(t)|^2] = \sum_{j=0}^{m} \alpha_m(j)R(j) \qquad (2.11)$$

Because of our definition of a model of a time series Y
as a decomposition into predictable and unpredictable parts,
prediction theory not only yields answers to the prediction
problem, which are of interest in themselves, but also helps
define types of models to be considered for stationary time
series model fitting. The usual autoregressive-moving average
definitions of time series schemes (model types) in reality are
formulas for the infinite memory one-step ahead predictor
$Y^{\mu}(t)$.

<u>Autoregressive scheme of order p</u> or AR(p):

$$-Y^{\mu}(t) = \alpha_p(1)Y(t - 1) + \ldots + \alpha_p(p)Y(t - p) \qquad (2.12)$$

<u>Moving average scheme of order q</u> or MA(q):

$$Y^{\mu}(t) = \beta_q(1)Y^{\nu}(t - 1) + \ldots + \beta_q(q)Y^{\nu}(t - q) \qquad (2.13)$$

<u>Mixed autoregressive-moving average scheme</u> or ARMA(p,q):

$$Y^{\mu}(t) = - [\alpha_{p,q}(1)Y(t - 1) + \ldots + \alpha_{p,q}(p)Y(t-p)]$$

$$+ \beta_{p,q}(1)Y^{\nu}(t - 1) + \ldots + \beta_{p,q}(q)Y^{\nu}(t - q)$$

$$(2.14)$$

These models are usually defined in terms of a *white noise*
time series $\varepsilon(t)$:

$$E[\epsilon(t)\epsilon(t + v)] = \sigma^2 I(v) \tag{2.15}$$

where I is the identity covariance defined by

$$I(v) = \begin{cases} 1 & \text{if } v = 0 \\ 0 & \text{if } v \neq 0 \end{cases} \tag{2.16}$$

One can show that $\epsilon(t) = Y^v(t)$, the innovation in the time
series at time t . In the sequel we will use both $\epsilon(t)$ and
$Y^v(t)$ to denote the true innovation in the time series.

3.   TRANSFER FUNCTIONS AND PREDICTION

From a finite sample (of size T) only a finite number of
parameters can be estimated. It is therefore attractive to fit
an observed time series by an ARMA (p,q) model where the orders
p and q are chosen as low as possible. Such a model has
parameters

$$\alpha_{p,q}(1), \ldots, \alpha_{p,q}(p), \beta_{p,q}(1), \ldots, \beta_{p,q}(q), \sigma^2_{p,q}$$

$$= E[Y^v(t)] \tag{3.1}$$

There is an extensive literature on how to estimate efficiently
these parameters (parameter identification) when the orders p
and q are known. Determining the orders p and q (model
identification) has also received much attention (notably,
Box and Jenkins (1970)). I would like to describe my approach
which does not postulate that there exists *true* orders p and
q and thus a *true* finite parameter model to be determined;
rather any finite parameter model is an *approximator* to *the
true infinite parameter model*. To make this assertion precise, it is useful to rewrite our models in operator form.
Let L denote the backward shift operator (denoted by B in
Box and Jenkins (1970)):

$$LY(t) = Y(t - 1) \tag{3.2}$$

$$L\epsilon(t) = \epsilon(t - 1) \tag{3.3}$$

$$L^p Y(t) = Y(t - p) \tag{3.4}$$

$$L^q \varepsilon(t) = \varepsilon(t - q) \tag{3.5}$$

Define transfer functions

$$g_p(L) = 1 + \alpha_p(1)L + \ldots + \alpha_p(p)L^p \tag{3.6}$$

$$h_q(L) = 1 + \beta_q(1)L + \ldots + \beta_q(q)L^q \tag{3.7}$$

in terms of which the ARMA (p,q) model can be written

$$g_p(L)Y(t) = h_q(L)\varepsilon(t) \tag{3.8}$$

Now under reasonable regularity conditions (say, continuous nonvanishing spectral density function), a stationary time series has an *infinite autoregressive representation*

$$g_\infty(L)Y(t) = \varepsilon(t) \tag{3.9}$$

and an *infinite moving average representation*

$$Y(t) = h_\infty(L)\varepsilon(t) \tag{3.10}$$

These transfer functions are assumed to be invertible; then

$$g_\infty(L) = h_\infty^{-1}(L) = h_q^{-1}(L)g_p(L) \tag{3.11}$$

This relation is fundamental because of the basic theoretical role of the infinite autoregressive and infinite moving average representations of a time series even when one postulates the existence of a finite parameter model.

Note that the AR(∞) and MA(∞) representations always exist while the AR(p,q) representation need not exist (and in my view is best regarded as an approximator to the infinite parameter models).

Given the AR(∞) and MA(∞) representations one can solve the k-step ahead prediction problem, which is to find, for $k = 1, 2, \ldots$,

$$Y^{\mu}(t + k\,|\,t) = E\,[\,Y(t + k)\,|\,Y(t),\ Y(t - 1),\ \ldots\ ] \qquad (3.12)$$

$$\sigma^2_k = E\,[\,|\,Y^{\nu}(t + k\,|\,t)\,|^2\,] \qquad (3.13)$$

$$Y^{\nu}(t + k\,|\,t) = Y(t + k) - Y^{\mu}(t + k\,|\,t) \qquad (3.14)$$

Using the MA(∞) representation one can write

$$Y(t + k) = \varepsilon(t + k) + \beta_{\infty}(1)\varepsilon(t + k - 1) + \ldots$$
$$+ \beta_{\infty}(k-1)\varepsilon(t + 1) + \beta_{\infty}(k)\varepsilon(t) + \ldots \qquad (3.15)$$

Take the conditional expectation with respect to $Y(t)$, $Y(t - 1)$, $\ldots$; then

$$Y^{\mu}(t + k\,|\,t) = \beta_{\infty}(k)\varepsilon(t) + \beta_{\infty}(k + 1)\varepsilon(t - 1) + \ldots \qquad (3.16)$$

$$Y^{\nu}(t + k\,|\,t) = \varepsilon(t + k) + \beta_{\infty}(1)\varepsilon(t + k - 1) + \ldots$$
$$+ \beta_{\infty}(k - 1)\,\varepsilon(t + 1) \qquad (3.17)$$

$$\sigma^2_k = \sigma^2\,[1 + \beta^2_{\infty}(1) + \ldots + \beta^2_{\infty}(k - 1)] \qquad (3.18)$$

The MA(∞) representation provides a formula for σ^2_k , and the AR(∞) representation provides a recursive formula for $Y^{\mu}(t + k\,|\,t)$, as follows.

$$-Y(t + k) = \alpha_{\infty}(1)Y(t + k - 1) + \ldots + \alpha_{\infty}(k - 1)Y(t + 1)$$
$$+ \alpha_{\infty}(k)Y(t) + \ldots \qquad (3.19)$$

$$-Y^{\mu}(t + k\,|\,t) = \alpha_{\infty}(1)Y^{\mu}(t + k - 1\,|\,t) + \ldots + \alpha_{\infty}(k - 1)Y^{\mu}(t + 1\,|\,t)$$
$$+ \alpha_{\infty}(k)Y(t) + \alpha_{\infty}(k + 1)Y(t - 1) + \ldots$$

$$(3.20)$$

4. EFFECT OF ESTIMATION AND MODEL MISSPECIFICATION ON PREDICTION

The foregoing formulas, simple as they seem, are conceptually extremely powerful and enable us to study the effects of parameter estimation, as well as incorrect model specification, on our predictions.

Let

$$g_p^*(L) = 1 + \alpha_p^*(1)L + \ldots + \alpha_p^*(p)L^p \tag{4.1}$$

$$h_q^*(L) = 1 + \beta_q^*(1)L + \ldots + \beta_q^*(q)L^q \tag{4.2}$$

be transfer functions (found either by estimation or by misspecification) such that one hopes that

$$\varepsilon^*(t) = h_q^{*-1}(L)g_p^*(L)Y(t) \tag{4.3}$$

is white noise. This model (called the *asterisk model*) is said to fit, or to be adequate, if the $\varepsilon^*(t)$ series (called the *asterisk innovations*) satisfies two basic conditions:

(1) $\varepsilon^*(t)$ is sufficiently close to white noise,

(2) $\sigma_{\varepsilon^*}^2 = E[|\varepsilon^*(t)|^2]$ is sufficiently close to σ_ε^2 ,

the true innovation variance.

An asterisk model has several important implications which play a role in answering the foregoing questions. First, it generates the following formula for the spectral density of the time series $Y(t)$.

$$f^*(\omega) = \frac{1}{2\pi}\sigma_{\varepsilon^*}^2 |h_q^*(e^{i\omega}) \div g_p^*(e^{i\omega})|^2 \tag{4.4}$$

Second, for k-step ahead predictors, it generates a recursive formula

$$-Y^{\mu*}(t + k|t) = \alpha_\infty^*(1)Y^{\mu*}(t + k - 1|t) + \ldots + \alpha_\infty^*(k$$
$$- 1) Y^{\mu*}(t + 1|t) + \alpha_\infty^*(k)Y(t)$$
$$+ \alpha_\infty^*(k + 1)Y(t - 1) + \ldots \tag{4.5}$$

Remarkably, for the difference $\delta_t(k)$ between the true k-step ahead predictor and the estimated k-step ahead predictor, we can obtain a formula

$$\delta_t(k) = Y^\mu(t + k \mid t) - Y^{\mu*}(t + k \mid t) \tag{4.6}$$

This formula is due to William S. Cleveland (1971). Write

$$\varepsilon^*(t + k) = Y(t + k) + \alpha_\infty^*(1)Y(t + k - 1) + \ldots \tag{4.7}$$

and take conditional expectation with respect to $Y(t)$, $Y(t - 1)$, $\ldots$, which is equivalent to conditional expectation with respect to $\varepsilon^*(t)$, $\varepsilon^*(t - 1)$, $\ldots$.

$$\varepsilon^{*\mu}(t + k \mid t) = Y^\mu(t + k \mid t) + \alpha_\infty^*(1)Y^\mu(t + k - 1 \mid t) + \ldots \tag{4.8}$$

Subtracting (4.5) from (4.8) we obtain

$$\varepsilon^{*\mu}(t + k \mid t) = \delta_t(k) + \alpha_\infty^*(1)\delta_t(k - 1) + \ldots + \alpha_\infty^*(k - 1)\delta_t(1) \tag{4.9}$$

Define $\delta_t(j) = 0$, $j = 0, -1, -2, \ldots$, and $\eta_t(k) = \varepsilon^{*\mu}(t + k \mid t)$, $k = 1, 2, \ldots$. Then we can write (4.9) in operator form,

$$\eta_t(k) = g_\infty^*(L)\delta_t(k), \quad k = 1, 2, \ldots \tag{4.9'}$$

A formula for $\delta_t(k)$ is then given by

$$\delta_t(k) = h_\infty^*(L)\eta_t(k) \tag{4.10a}$$

or, explicitly,

$$Y^\mu(t + k \mid t) - Y^{\mu*}(t + k \mid t) = \sum_{j=0}^{k-1} \beta_\infty^*(j)\varepsilon^{*\mu}(t + k - j \mid t) \tag{4.10b}$$

The question of how close $\varepsilon^*(t)$ is to white noise is a question of how true it is that

$$\varepsilon^{*\mu}(t + k \mid t) = 0 \ , \quad k = 1, 2, \ldots, h$$

where h is the horizon for which we desire to predict. Another use of (4.10), not yet explored, is to improve a predictor by predicting its asterisk innovations.

The question of what determines how close $\sigma^2_{\varepsilon^*}$ is to σ^2_{ε} has been studied by Grenander and Rosenblatt (1957) who show (p. 269) that approximately

$$\sigma^2_{\varepsilon^*} = \sigma^2_{\varepsilon} \ (1 + \frac{1}{2} \ J) \tag{4.11}$$

where

$$J = \frac{1}{2\pi} \int_{-\pi}^{\pi} \left(\frac{f(\omega) - f^*(\omega)}{f(\omega)} \right)^2 d\omega - \left(\frac{1}{2\pi} \int_{-\pi}^{\pi} \frac{f(\omega) - f^*(\omega)}{f(\omega)} \ d\omega \right)^2 \tag{4.12}$$

We see that $\sigma^2_{\varepsilon^*}$ is close to σ^2_{ε} if and only if J is close to zero. From J we get a precise sense in which the approximating spectral density $f^*(\omega)$ is close to the true spectral density $f(\omega)$ when $\sigma^2_{\varepsilon^*}$ is close to σ^2_{ε} .

It seems clear then that one important measure of a finite parameter model is how close $\sigma^2_{\varepsilon^*}$ (its mean square one-step prediction error) comes to σ^2_{ε} . It seems desirable to directly estimate σ^2_{ε} , and one way to do this is via the theoretical expression

$$\log \ \sigma^2_{\varepsilon} = \frac{1}{2\pi} \int_{-\pi}^{\pi} \log \ 2\pi f(\omega) \, d\omega \tag{4.13}$$

(See Davis and Jones (1968) for an estimator of $\log \ \sigma^2_{\varepsilon}$.)

5. CRITERION FOR CHOOSING THE ORDER OF A BEST-FITTING
AUTOREGRESSIVE SCHEME

The approach to stationary time series modeling that I
propose is that the *true* model of a time series is AR(∞), an
infinite order autoregressive scheme, whose transfer function
$g_\infty(L)$ we desire to estimate. The ARMA models (considered as
the universal model by some analysts) are introduced only to
provide a representation of $g_\infty(L)$ as a function of a finite
number of parameters, and thereby reduce the estimation of
$g_\infty(L)$ to a customary parameter estimation problem. I would
like to show optimally how to estimate $g_\infty(L)$ *nonparametrically*.
One then has a nonparametric solution of the prediction problem.
For model identification, one can seek to find various finite
parameter transfer functions $h_q^{-1}(L)g_p(L)$ close to $g_\infty(L)$.

To estimate $g_\infty(L)$ it is natural to adopt the penalty
function $E[J]$ as the criterion for an estimator $\hat{g}_\infty(L)$ where

$$J = \frac{1}{2\pi} \int_{-\pi}^{\pi} \left| \frac{\hat{g}_\infty(e^{i\omega}) - g_\infty(e^{i\omega})}{g_\infty(e^{i\omega})} \right|^2 d\omega \tag{5.1}$$

The corresponding spectral density estimator $\hat{f}(\omega)$ will then
be a good estimator of the true spectral density $f(\omega)$ by the
criterion J introduced for spectral densities (in Section 4).

It is natural to consider first the nonstatistical problem
of approximating $g_\infty(L)$ by a polynomial

$$p_m(L) = a_0 + a_1 L + \ldots + a_m L^m \tag{5.2}$$

of degree m, chosen to be the closest such polynomial in the
distance function

$$J_m = \frac{1}{2\pi} \int_{-\pi}^{\pi} \left| \frac{p_m(e^{i\omega}) - g_\infty(e^{i\omega})}{g_\infty(e^{i\omega})} \right|^2 d\omega \tag{5.3}$$

One can show that J_m has minimum value

$$J_{m,min} = 1 - \frac{\sigma_\infty^2}{\sigma_m^2} \tag{5.4}$$

achieved at the polynomial

$$\frac{\sigma_\infty^2}{\sigma_m^2} \, g_m(e^{i\omega}) \tag{5.5}$$

where $g_m(L)$ is the transfer function of the optimum memory m predictor, and σ_m^2 is the mean square prediction error predicting one-step ahead with memory m . Of course the true covariance, and therefore the true transfer function, is unknown. However, one can form an estimated transfer function

$$\hat{g}_m(L) = \sum_{j=0}^{m} \hat{\alpha}_m(j) L^j \tag{5.6}$$

from the estimated autoregressive coefficients

$$\sum_{j=0}^{m} \hat{\alpha}_m(j) R_T(j - k) = 0 \, , \quad k = 1, \ldots, m \tag{5.7}$$

where $\hat{\alpha}_m(0) = 1$ and

$$R_T(v) = \frac{1}{T} \sum_{t=1}^{T-v} Y(t) Y(t + v) \tag{5.8}$$

is the sample covariance function. To estimate σ_m^2 one uses

$$\hat{\sigma}_m^2 = \sum_{j=0}^{m} \hat{\alpha}_m(j) R_T(j) \tag{5.9}$$

or the unbiased estimator $(T/T-m) \, \hat{\sigma}_m^2 = \hat{\hat{\sigma}}_m^2$.

Using the deep results proven by Kromer (1969) on the properties of $\hat{\alpha}_m(j)$ one can show that *approximately*

$$\frac{1}{2\pi} \int_{-\pi}^{\pi} E\left|\hat{g}_m(e^{i\omega}) - g_\infty(e^{i\omega})\right|^2 \left|g_\infty(e^{i\omega})\right|^{-2} d\omega = CAT(m) \quad (5.10)$$

defining

$$CAT(m) = 1 - \frac{\hat{\sigma}_\infty^2}{\hat{\sigma}_m^2} + \frac{m}{T} \quad (5.11)$$

We use the initials CAT to represent "criterion auto-regressive transfer function" to distinguish it from another interesting criterion [which I denote CIA(m), "criterion information Akaike"] defined by Akaike (1971) as

$$CIA(m) = \log \hat{\sigma}_m^2 + 2 \frac{m}{T}$$

The value $\hat{m}$ minimizing these criteria is chosen as the order of an autoregressive scheme chosen to "fit" the observed time series.

Examples of model-fitting real time series that we have tried indicate that these two criteria give almost identical results for the order $\hat{m}$ of the best-fitting scheme. Akaike interprets the meaning of the estimated order $\hat{m}$ as an es-timator of the true order m , while in my approach $\hat{g}_{\hat{m}}(L)$ is an estimator of $g_\infty(L)$. I call $g_\infty(L)$ the ARTF (auto-regressive transfer function) and $\hat{g}_{\hat{m}}(L)$ the ARTFACT (auto-regressive transfer function converging to the truth).

We finally return to the distinction between hard and soft rules. The criterion of minimizing CAT(m) is motivated by certain hard results to which we allude in the final section, but ultimately it is by a soft rule, whose definition in practice is guided by the results which it yields.

6. SOME HARD RULES

The criterion CAT(m) is composed of two parts:
$1 - \{\sigma_\infty^2 \div \sigma_m^2\}$ represents the bias due to using a polynomial

(a finite number of parameters) to approximate an infinite
series, while m/T represents the overall variance of estimating the m-parameter approximating polynomial.

   To derive the bias term we consider the problem of finding the polynomial p of degree m minimizing J_m . By the
projection theorem in Hilbert space, there is a unique minimizing polynomial p^* characterized by normal equations,

$$\int_{-\pi}^{\pi} p^*(e^{i\omega}) e^{-ik\omega} \left| g_\infty(e^{i\omega}) \right|^{-2} d\omega = \int_{-\pi}^{\pi} g_\infty(e^{i\omega}) e^{-ik\omega} \left| g_\infty(e^{i\omega}) \right|^{-2} d\omega$$

(6.1)

for k = 0, 1, ..., m, since any m-degree polynomial $p(e^{i\omega})$
is a linear combination of 1, $e^{i\omega}$, ..., $e^{im\omega}$.

   Now the transfer function $g_m(e^{i\omega})$ of best memory m predictor satisfies the normal equations

$$\int_{-\pi}^{\pi} g_m(e^{i\omega}) e^{-ik\omega} f(\omega) = \begin{cases} \sigma_m^2 , & k = 0 \\ \\ 0 , & k = 1, 2, ..., m \end{cases}$$

(6.2)

Since

$$f(\omega) = \frac{1}{2\pi} \sigma_\infty^2 \left| g_\infty(e^{i\omega}) \right|^{-2}$$

(6.3)

it follows from (6.2) that

$$p^* = (\sigma_\infty^2 / \sigma_m^2) g_m$$

(6.4)

is the solution of (6.1).

   The normal equations (6.1) imply that $p^* - g_\infty$ and p^*
are orthogonal; therefore (dropping all reference to the
variable of integration ω)

$$\frac{1}{2\pi} \int_{-\pi}^{\pi} \left| p^* - g_\infty \right|^2 \left| g_\infty \right|^{-2} = \frac{1}{2\pi} \int_{-\pi}^{\pi} \left| g_\infty \right|^2 \left| g_\infty \right|^{-2}$$

(6.5)

$$- \frac{1}{2\pi} \int_{-\pi}^{\pi} |p^*|^2 |g_\infty|^{-2} = 1 - \frac{\sigma_\infty^2}{\sigma_m^4} \int_{-\pi}^{\pi} |g_m|^2 f = 1 - \frac{\sigma_\infty^2}{\sigma_m^2} \qquad (6.6)$$

This completes the proof of our expression for bias.

The expression for variance can be motivated intuitively as follows. Essentially, variance equals

$$\frac{1}{2\pi} \int_{-\pi}^{\pi} \mathrm{var}(\hat{g}_m) |g_\infty|^{-2} \qquad (6.7)$$

which equals m/T for large m under the approximation (see Kromer (1969) and Parzen (1969))

$$\mathrm{Var}[\hat{g}_m(e^{i\omega})] = \frac{m}{T} |g_\infty(e^{i\omega})|^2 \qquad (6.8)$$

More precise bounds for variance can be given using the results in Kromer's thesis (1969).

BIBLIOGRAPHY

Akaike, H. (1971). Autoregressive model fitting for
control. *Ann. Inst. Statist. Math.* **23**, 163-80.

Akaike, H. (1971). Information theory and an extension
of the maximum likelihood principle. Presented at
the Second International Symposium on Information
Theory, Tsahkadsor, Armenian SSR, 2-8 September,
1971. To appear in *Problems of Control and
Information Theory*, Akademiai Kiado (Publishing
House of the Hungarian Academy of Sciences,
Budapest 502, P. O. Box 24, Hungary).

Akaike, H. (1972). Use of an information theoretic
quantity for statistical model identification.
Proc. 5th Hawaii Int. Conf. System Sci., Western
Periodicals Co., 249-50.

Box, G. E. P. and G. M. Jenkins (1970). *Time Series,*
Forecasting, and Control. San Francisco:
Holden-Day.

Cleveland, W. S. (1971). Fitting time series models
for prediction. *Technometrics* 13, 713-23.

Davis, H. T. and R. H. Jones (1968). Estimation of the
innovation variance of a stationary time series, *J. Amer.*
Statist. Assoc. 63, 141-48.

Grenander, U. and M. Rosenblatt (1957). *Statistical*
Analysis of Stationary Time Series. New York:
John Wiley and Sons, Inc.

Kromer, R. E. (1969). "Asymptotic Properties of the
Autoregressive Spectral Estimator", Ph.D. Thesis,
Department of Statistics, Stanford University.

Parzen, E. (1969). Multiple time series modeling,
Multivariate Analysis, Vol. II, (ed. P. R. Krishnaiah)
New York: Academic Press, 389-410.

CROSSING PROBABILITY BOUNDS FOR NONSTATIONARY
GAUSSIAN PROCESSES

L. H. Koopmans and Clifford Qualls

The University of New Mexico
Albuquerque, New Mexico

1. INTRODUCTION

We are interested in the probability that a real-valued
second order nonstationary Gaussian process $\{Y(t): t \in T\}$
with $E(Y(t)) = 0$ <u>ever</u> crosses levels $\pm a$. We restrict
attention to the situation in which the index set T is the
whole real line. However, what we discuss is also valid when
T is a multidimensional set. Thus, in particular the results
are valid for random scalar fields.

It is assumed that $Y(t)$ is the output of a linear system
with known impulse response function $h(t)$ to an input process
$X(t)$, where $\int h^2(t)dt = N^2 < \infty$. This means that

$$Y(t) = \int h(t-\tau)X(\tau)d\tau \tag{1.1}$$

Alternatively, in the case of random fields where T is
a two- or three-dimensional spatial domain of points $\underset{\sim}{t}$, it
is assumed that the field $X(\underset{\sim}{t})$ is observed by means of probe
functions $h_p(\underset{\sim}{t})$ centered at points $\underset{\sim}{p}$ in T for which
$\int h_p^2(\underset{\sim}{t})d\underset{\sim}{t} \leq N^2 < \infty$ for all $\underset{\sim}{p}$. Then, what is observed are
the output values

$$Y_p = \int h_p(\underset{\sim}{t})X(\underset{\sim}{t})d\underset{\sim}{t} \tag{1.2}$$

Our interest, then, is in the probability that Y_p crosses
given levels $\pm a$ at any point $\underset{\sim}{p}$ in the field.

The probabilities of interest, called crossing probabilities,
play an interesting and important role in many applied problems.
Some examples are given next.

2. PROBLEMS IN WHICH CROSSING PROBABILITIES ARE OF INTEREST

The first type of problem falls in the general category
known as detection problems. It is of interest to determine
the presence of an object in a very noisy environment. If the
object is present, it will increase the amplitude of the random
process representing the noise environment at some point in
time or space. An example is the detection of a distant air-
plane by radar. Another is the electromagnetic detection of a
submarine in the ocean. We would try to establish a level such
that the random process crosses the level if the signal is
present. It must be set high enough to make the probability of
crossing small if no signal is present, but reasonably large
when it is. Thus, we are faced with the problem of evaluating--
or bounding--the probability $P(|Y(t)| > a$ for some $t)$ as a
function of a.

The second type of problem arises as a result of defining
a mechanism of failure in terms of level crossings. We describe two such problems. Howell and Lin (1971) introduce a
nonstationary model for atmospheric turbulence to study the
response characteristics of flight vehicles. By using a nonstationary process they model the kinds of variations in gust
velocities which one might encounter in low-level flight over
rough terrain. When the response (velocity or acceleration) of
the vehicle (the linear system) exceeds a certain level the "ride"
may be classified as "intolerable" for passenger vehicles and
attempts would be made to design the vehicle to minimize the

probability of exceeding this level in anticipated turbulent environments. In the case of unmanned vehicles such as missiles, the vehicle may be damaged or destroyed if response accelerations exceed a certain level, and again, it would be desirable to design the vehicle to minimize the probability of such occurrences.

In quite another context (Koopmans, et al. (1973)), we hit upon the same model for nonstationary processes that Howell and Lin used. The problem is one of modeling earthquake environments to study the failure of buildings, bridges, etc., in the presence of strong motion earthquakes. The physical structure is modeled by the linear filter with impulse response function $h(t)$, and $X(t)$ represents the horizontal ground acceleration at the structure perpendicular to the line through the quake epicenter. The output process $Y(t)$ represents the displacement or "sway" of the structure and a reasonable model for failure is to say that the structure fails if for some t, $|Y(t)| > a$, where the level a depends on structural strength, etc. Then the failure probability is precisely the probability $P(|Y(t)| > a$ for some $t)$.

In general, even when $X(t)$ is Gaussian it is impossible to evaluate these probabilities, although asymptotic expressions for large a exist. Another alternative is to establish "tight" upper bounds on these probabilities. Previously published bounds (e.g., Pickands (1969)) are quite complicated and, moreover, only hold asymptotically as a tends to infinity. We have obtained an exponential bound which is valid for all interesting values of a, is very simple to apply in practice and is asymptotically of the same form as the more elaborate bounds established earlier. For these reasons, we feel that this bound will have considerable utility in studying the kinds of problems outlined above.

We were introduced to this problem by the earthquake engineer, J. T. P. Yao, currently Professor of Civil Engineering at Purdue University. The three of us publish jointly in Koopmans, et al. (1973) the results of a numerical application of our basic exponential bound for crossing probabilities to an earthquake engineering study. In Koopmans and Qualls (1972) we generalize the conditions under which the bound is valid and establish its correctness in what we believe to be the proper mathematical setting. The present paper is intended to be primarily an expository introduction to this theory. However, we also include some refinements and further generalizations which have been obtained recently.

3. THE BOUND FOR CROSSING PROBABILITIES

We show in this section that the following bound for crossing probabilities is obtained in a simple way from a corresponding bound for tail probabilities of the energy $M^2 = \int X^2(t)\,dt$ of the input process.

$$P\left(\left|Y(t)\right| > a \text{ for some } t\right) \leq \sqrt{1 + \frac{\varepsilon(a)}{\Lambda}}\ e^{-[\varepsilon(a)/2\Lambda]}$$

$$(3.1)$$

where $\varepsilon(a) = (a^2 - N^2 EM^2)/N^2$, $\Lambda^2 = \frac{1}{2} \text{Var } M^2$ and N^2 is the quantity defined in Section 1.

By the Cauchy-Schwarz inequality we have

$$\left|Y(t)\right| = \left|\int h(t-\tau)X(\tau)\,d\tau\right| \leq \sqrt{\int h^2(t-\tau)\,d\tau}\ \sqrt{\int X^2(\tau)\,d\tau} = MN$$

$$(3.2)$$

for all t. Consequently, the event $[\left|Y(t)\right| > a$ for some $t]$ is contained in the event $[MN > a] = [M^2 > (a^2/N^2)]$ $= [M^2 - EM^2 > (a^2 - N^2 EM^2/N^2)]$. It follows that

$$P(|Y(t)| > a \text{ for some } t) \leq P\left(M^2 - EM^2 > \frac{a^2 - N^2 EM^2}{N^2}\right) \quad (3.3)$$

Thus, (3.1) will be a consequence of the inequality

$$P(M^2 - EM^2 \geq \varepsilon) \leq \sqrt{1 + \frac{\varepsilon}{\Lambda}} \; e^{-(\varepsilon/2\Lambda)}, \; \varepsilon \geq 0 \quad (3.4)$$

To establish (3.4), we will require a special model for the nonstationary Gaussian input process $X(t)$. We develop this next.

4. THE MODEL FOR X(t)

We are led to the "correct" model for $X(t)$ by mimicking the process by which Professor Yao has created strong motion earthquakes on an analog computer for the purpose of estimating crossing probabilities by Monte Carlo simulation. A wide-band stationary process $U(t)$ is run through a linear filter with transfer function $G(\lambda)$ which excludes all frequencies except those characteristic of observed strong motion earthquakes. A tapering function $B(t)$ is then applied to the resulting process to produce the typical transient shape of an earthquake accelerogram. (See Figure 1).

Mathematically, the input process can be conveniently represented by the spectral decomposition for a weakly stationary process:

$$U(t) = \int e^{i\lambda t} Z(d\lambda) \quad (4.1)$$

where $Z(A)$ is an orthogonal random process with $EZ(A) = 0$ for all A. Note that if $U(t)$ is Gaussian, so is $Z(A)$. If $F(A)$ denotes the spectral distribution of $U(t)$, which satisfies the relationship

$$EU(t+\tau)U(t) = \int e^{i\tau\lambda} F(d\lambda) \quad (4.2)$$

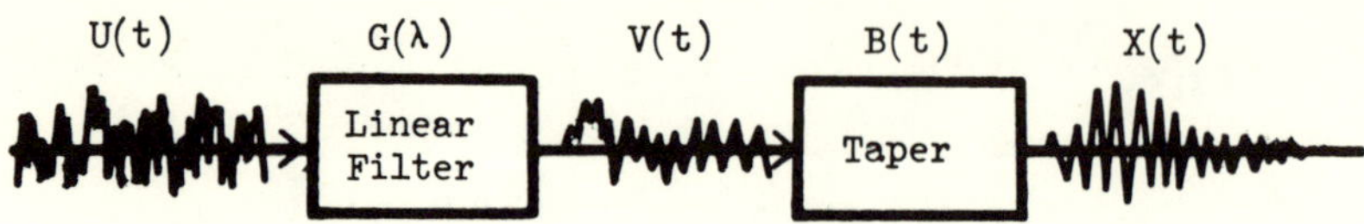

Figure 1

then $\int F(d\lambda) < \infty$ and the above process is related to this function by the expression

$$EZ(A)\overline{Z(B)} = F(A \cap B) \tag{4.3}$$

In particular, if $A \cap B = \phi$ we obtain the orthogonality relation

$$EZ(A)\overline{Z(B)} = 0 \tag{4.4}$$

The result of filtering $U(t)$ is a weakly stationary process $V(t)$ with spectral representation

$$V(t) = \int G(\lambda)e^{i\lambda t}Z(d\lambda) \tag{4.5}$$

Next, the tapering function multiplies $V(t)$ to yield

$$X(t) = \int B(t)G(\lambda)e^{i\lambda t}Z(d\lambda) \tag{4.6}$$

It is assumed that the tapering function is square integrable and necessarily $G(\lambda)$ is square integrable with respect to F in order for $V(t)$ to be a valid weakly stationary process. Consequently, the process $X(t)$ has a representation of the form

$$X(t) = \int B(t,\lambda)Z(d\lambda) \tag{4.7}$$

with

$$\int\int |B(t,\lambda)|^2 F(d\lambda)dt < \infty \tag{4.8}$$

In order for $X(t)$ to be real-valued, it is also necessary that

$$B(t,-\lambda) = \overline{B(t,\lambda)} \tag{4.9}$$

Our results will be valid for any process of the form (4.7) subject to the condition (4.8). Because (4.8) implies that the process is transient, it is decidedly nonstationary.

Without the integral over t in (4.8), this type of process has long been one of the more popular models for non-stationary processes. The work of M. B. Priestley and his colleagues at University of Manchester (England) has been particularly significant in advancing the importance of this model (see Priestley (1965)).

Another important model for a stochastic process of the form (4.7) is that of the response to a time-varying linear filter with a given weakly stationary stochastic input process $\{\varepsilon(t): t \in T\}$. The linear filter is described by a linear differential operator with time-varying coefficients; $\sum_{j=0}^{n} a_j(t)D^j$, where $D = \frac{d}{dt}$. Thus, the process $\{X(t): t \in T\}$ can be defined implicitly as a particular solution to the differential equation

$$\sum_{j=0}^{n} a_j(t)D^j X(t) = \varepsilon(t) \tag{4.10}$$

It is now possible to write $X(t)$ explicitly in the form

$$X(t) = \int B(t,\lambda) Z_\varepsilon(d\lambda) \tag{4.11}$$

where

$$B(t,\lambda) = \frac{e^{i\lambda t}}{\sum_{j=0}^{n} a_j(t)(i\lambda)^j} \tag{4.12}$$

and $Z_\varepsilon(A)$ is the orthogonal random process appearing in the spectral representation of $\varepsilon(t)$. Conditions on the roots of the time-varying characteristic equation

$$\sum_{j=0}^{n} a_j(t) z^j = 0 \tag{4.13}$$

can be imposed to ensure (3.8), where $F(A) = F_\varepsilon(A)$ is the spectral distribution of $\varepsilon(t)$. If, for example, $\varepsilon(t)$ is taken to be white noise with unit variance, the model is completely determined by the coefficients $a_j(t)$ and by varying these functions a large class of input processes $X(t)$ can be produced.

5. THE PROBABILITY BOUND FOR THE ENERGY OF THE PROCESS

We assume that the process (4.7) is Gaussian in this section. Our earlier proof of the inequality (3.4) can be somewhat simplified as was indicated to us by R. Wijsman:

Let

$$K(x,y) = \int B(t,x)\overline{B(t,y)}\,dt . \tag{5.1}$$

By straightforward computations we can show that $K(x,y)$ is nonnegative definite and

$$\int\int |K(x,y)|^2 F(dx)F(dy) \le \left[\int\int |B(t,x)|^2 F(dx)dt \right]^2 < \infty . \tag{5.2}$$

This implies that $K(x,y)$ is a Hilbert-Schmidt kernel (Riesz and Sz.-Nagy (1955), p. 242). Hence it has an $L_2(F)$ convergent expansion,

$$K(x,y) = \sum_{j=1}^{\infty} \lambda_j \phi_j(x)\overline{\phi_j(y)} \tag{5.3}$$

in terms of the eigenvalues and eigenfunctions of the associated operator. The λ_j's are real and nonnegative and the ϕ_j's are orthonormal in $L_2(F)$;

$$\int \phi_j(x)\overline{\phi_k(x)} F(dx) = \begin{cases} 1, & j = k \\ 0, & j \ne k \end{cases} \tag{5.4}$$

Now, let

$$W_i = \int \phi_i(x) Z(dx), \quad i = 1, 2, \ldots \tag{5.5}$$

Since $Z(A)$ is Gaussian, the W_i's are normal random variables. It is easily seen that the condition $B(t,-\lambda) = \overline{B(t,\lambda)}$ implies $\phi_i(-x) = \overline{\phi_i(x)}$. Then, by the change of variables $y = -x$,

$$\overline{W_i} = \int_{-\infty}^{\infty} \overline{\phi_i(x)} \ \overline{Z(dx)}$$

$$= \int_{-\infty}^{\infty} \phi_i(-x) Z(-dx)$$

$$= \int_{-\infty}^{\infty} \phi_i(y) Z(dy)$$

$$= W_i \tag{5.6}$$

and it follows that W_i is real-valued. Moreover, $EW_i = 0$ and

$$EW_i W_j = \int \phi_i(x) \overline{\phi_j(x)} \ F(dx) \tag{5.7}$$

Thus, by (5.4), the W_i's are uncorrelated and have variance 1.

Finally, because of (4.7),

$$M^2 = \int X^2(t) dt = \int \int K(x,y) Z(dx) \overline{Z(dy)}$$

$$= \sum_{j=1}^{\infty} \lambda_j \int \int \phi_j(x) \overline{\phi_j(y)} Z(dx) \overline{Z(dy)}$$

$$= \sum_{j=1}^{\infty} \lambda_j W_i^2 \tag{5.8}$$

This is a representation of M^2 as a linear combination of independent chi-square variables, each with one degree of freedom. It follows that

$$EM^2 = \sum_{j=1}^{\infty} \lambda_j \tag{5.9}$$

and

$$\mathrm{Var}\ M^2 = \sum_{j=1}^{\infty} \lambda_j^2\ \mathrm{Var}\ W_j^2$$

$$= 2 \sum_{j=1}^{\infty} \lambda_j^2 \tag{5.10}$$

Let $U_j = W_j^2 - 1$. Then $M^2 - EM^2 = \sum_{j=1}^{\infty} \lambda_j U_j$. The moment generating function of U_j is

$$E\ \exp(tU_j) = \frac{e^{-t}}{(1-2t)^{1/2}}\ ,\quad t < \frac{1}{2} \tag{5.11}$$

The function $C(t)$ defined implicitly by the expression

$$\exp[C(t)t^2] = \frac{e^{-t}}{(1-2t)^{1/2}} \tag{5.12}$$

for $0 < t < \frac{1}{2}$, is easily seen to be increasing in t over this range. Taking $\Lambda^2 = \sum_{j=1}^{\infty} \lambda_j^2$, it follows that for all j,

$$C(\theta\lambda_j) \leq C(\theta\Lambda)\ ,\quad 0 < \theta < \frac{1}{2\Lambda} \tag{5.13}$$

Now, by a well-known probability inequality (Loève (1963), p. 255), for $0 < \theta < \frac{1}{2\Lambda}$,

$$P(M^2 - EM^2 \geq \varepsilon) \leqq E\ \exp\left[\theta\left(\sum_{j=1}^{\infty} \lambda_j U_j - \varepsilon\right)\right]$$

$$= e^{-\varepsilon\theta}\ \prod_{j=1}^{\infty} E\ \exp(\theta\lambda_j U_j)$$

$$= e^{-\varepsilon\theta}\ \exp\left[\theta^2 \sum_{j=1}^{\infty} C(\theta\lambda_j)\lambda_j^2\right]$$

$$\leqq e^{-\varepsilon\theta}\ \exp[C(\theta\Lambda)\theta^2\Lambda^2]$$

$$= e^{-\varepsilon\theta}\ \frac{e^{-\theta\Lambda}}{(1-2\theta\Lambda)^{1/2}} \tag{5.14}$$

By minimizing this expression as a function of θ , the bound (3.4) is obtained.

For the purpose of using the crossing probability bound (3.1) in applications, it is convenient to have the parameters EM^2 and Λ expressed in terms of the functions appearing in the model representation (4.7) and (4.8). By direct computation it is easily seen that

$$EM^2 = \int \int |B(t,\lambda)|^2 F(d\lambda) dt \qquad (5.15)$$

Moreover, we find that

$$\Lambda^2 = \frac{1}{2} \operatorname{Var} M^2 = \frac{1}{2} \int \int |K(x,y)|^2 F(dx) F(dy) \qquad (5.16)$$

where $K(x,y)$ is defined by (5.1).

<u>Example 5.1.</u> In the earthquake study (Koopmans, et al. (1973) we take the linear filter with transfer function $G(\lambda)$ (see Figure 1) to be an ideal band pass filter;

$$G(\lambda) = \begin{cases} 1, & \lambda_1 < |\lambda| < \lambda_2 , \quad (\lambda_1 > 0) \\ 0, & \text{otherwise} \end{cases} \qquad (5.17)$$

and the input process to be normalized white noise over the pass band of this filter;

$$F(d\lambda) = \begin{cases} cd\lambda, & \lambda_1 < |\lambda| < \lambda_2 \\ 0 , & \text{otherwise} \end{cases} \qquad (5.18)$$

where

$$c = \frac{1}{2(\lambda_2 - \lambda_1)} \qquad (5.19)$$

The tapering function is taken to be of the form

$$B(t) = \begin{cases} g(e^{-\alpha t} - e^{-\beta t}), & t > 0 \\ 0, & \text{otherwise} \end{cases} \qquad (5.20)$$

With $\lambda_1 = 0.6$ rad., $\lambda_2 = 13.6$ rad., $\alpha = 0.085$, $\beta = 0.170$ and $g = 10$, the sample functions correspond roughly to earthquakes of magnitude 7 on the Richter scale.

We analyze the failure probabilities for a simple structure with impulse response function

$$h(t) = \begin{cases} ke^{-\rho_0 t} \sin \lambda_d t \, , \ t > 0 \\ 0, \ t \leq 0 \end{cases} \qquad (5.21)$$

where $\lambda_d = \lambda_0 \sqrt{1-\rho_0^2}$ and k is a normalizing constant. This is the impulse response of a second order system with fundamental frequency λ_0 and damping factor ρ_0 . Curves giving the upper bounds on the probability of failure as a function of the crossing level a are provided in Koopmans, et al. (1973).

In the next section, some miscellaneous extensions of the above results are given.

6.   EXTENSIONS

6.1.   Curve Crossing Probability Bounds

It is sometimes of interest to have a bound for probabilities of the form

$$P(|Y(t)| \geq ar(t) \quad \text{for some} \quad t) \qquad (6.1)$$

where $r(t)$ is a positive function. Thus, in the earthquake engineering example it would be possible to include fatigue of the structural materials in the failure mechanism. A structure which has been stressed near its failure limit during the early part of an earthquake fails at lower stress levels later. This can be accounted for by allowing $r(t)$ to decrease with time to a lower failure level.

If $r(t)$ is given in advance, this modification can be easily accounted for as follows. Note that the probability (6.1) is the same as the probability $P(|W(t)| \geq a$ for some t), where $W(t) = Y(t)/r(t)$. But then $W(t) = \int g(t,\tau)X(\tau)d\tau$ where $g(t,\tau) = h(t-\tau)/r(t)$. Now, as long as $\int g^2(t,\tau)d\tau \leq N^2$ for all t, the bound (3.1) will hold for (6.1) with this value of N^2 .

This analysis also suggests that the fatigue mechanism can be built into the impulse response function of the structure to produce a time dependent impulse response $h(t,\tau)$. Then, as long as $\int h^2(t,\tau)d\tau \leq N^2$ for all t, the bound (3.1) can be used.

6.2. A "Model-Free" Crossing Probability Bound

Let $X(t)$ be a second order Gaussian process with zero mean function and covariance function

$$R(t,s) = EX(t)X(s) \tag{6.2}$$

It is possible to obtain the bound (3.1) for.

$Y(t) = \int h(t-\tau)X(\tau)d\tau$ without explicitly introducing the model (4.7). The bound gains somewhat from a theoretical viewpoint, since the conditions under which it holds are different from the conditions which have been obtained previously. However, it also loses somewhat in practical utility.

It is to be expected that stochastic processes of interest in crossing problems will have some form of continuity. Let $L_2(X)$ denote the set of all random variables which are either finite linear combinations of the form $\sum_j c_j X(t_j)$ or limits in mean-square of Cauchy-sequences of such linear combinations. Then the process is said to be weakly continuous if the function $g_U(t) = E(UX(t))$ is continuous for every $U \in L_2(X)$. Our theorem follows:

<u>Theorem 6.2.1.</u> If $X(t)$ is weakly continuous and $\int R(t,t)dt < \infty$, then (3.1) is valid for $Y(t) = \int h(t-\tau)X(\tau)d\tau$, where

$$\int h^2(t)dt = N^2 , \quad EM^2 = \int R(t,t)dt \tag{6.3}$$

$$\Lambda^2 = \frac{1}{2} \text{Var } M^2 = \frac{1}{2} \int \int R^2(s,t)dsdt \tag{6.4}$$

<u>Proof.</u> The weak continuity of $X(t)$ is sufficient to establish the separability of $L_2(X)$ (e.g., see Parzen (1959),

Theorem 20). The reproducing kernel Hilbert space for which $R(s,t)$ is the reproducing kernel is also separable, since it is topologically equivalent to $L_2(X)$. A result of Aronszajn ((1950), pp. 368-371) is that the reproducing kernel can then be written in the form

$$R(s,t) = \int B(t,\lambda)\overline{B(s,\lambda)}V(d\lambda) \tag{6.5}$$

with

$$\int |B(t,\lambda)|^2 V(d\lambda) < \infty \tag{6.6}$$

The measure $V(A)$ is finite. But in Parzen (1959), Theorem 4B, it is shown that this is sufficient to guarantee the existence of an orthogonal set function $Z(B)$ for which $V(A)$ is the spectral measure; i.e.,

$$EZ(A)\overline{Z(B)} = V(A \cap B) \tag{6.7}$$

Moreover, the representation

$$X(t) = \int B(t,\lambda)Z(d\lambda) \tag{6.8}$$

is valid. Also, by virtue of (6.5), (6.6) and the condition $\int R(t,t)dt < \infty$, the basic model assumption (4.8) is satisfied. The bound now follows as before.

By the Fubini theorem,

$$EM^2 = \int EX^2(t)dt = \int R(t,t)dt \tag{6.9}$$

Another application of this theorem yields

$$\int \int R^2(x,t)dsdt = \int \int |K(x,y)|^2 V(dx)V(dy) \tag{6.10}$$

where $K(s,y)$ is defined in (5.1).

This establishes the expression for Λ^2.

BIBLIOGRAPHY

Aronszajn, N. (1950). Theory of reproducing kernels, *Trans. Amer. Math. Soc.* <u>68</u>, 337-404.

Howell, L. J. and Y. K. Lin (1971). Response of flight vehicles to nonstationary atmospheric turbulence, *Proc. AIAA/ASME 12th Structures, Structural Dynamics and Materials Conference, Anaheim, California.*

Koopmans, L. H. and C. Qualls (1972). An exponential probability bound for the energy of a type of Gaussian process, *Ann. Math. Statist.* <u>43</u>, 1953-60.

Koopmans, L. H., Qualls, C. and J. T. P. Yao (1973). An upper bound on the failure probability for linear structures. *J. Appl. Mechanics* (Ser. E) <u>40</u>, 181-5.

Loève, M. (1963). *Probability Theory,* (3rd ed.). Princeton: Van Nostrand.

Parzen, E. (1959). "Statistical inference on time series by Hilbert space methods, I," Stanford Univ. Dept. of Statistics Tech. Report 23, Reprinted in *Time Series Analysis Papers* (1967). San Francisco: Holden-Day.

Pickands, J., III (1969). Upcrossing probabilities for stationary Gaussian processes. *Trans. Amer. Math. Soc.* <u>145</u>, 51-73.

Priestley, M. B. (1965). Evolutionary spectra and nonstationary processes. *J. R. Statist. Soc.,* (Ser. B.), <u>27</u>, 204-37.

Riesz, F. and B. Sz.-Nagy (1955). *Functional Analysis.* New York: Unger.

MULTIVARIATE PREDICTION AND SPLINE FUNCTIONS

Herbert T. Davis

University of New Mexico
Albuquerque, New Mexico

1. INTRODUCTION

Given a sequence of times $\{t_1, \ldots, t_n\}$ in some time interval T $(=[0,1]$, say), and a set of values $\{X(t_1), \ldots, X(t_n)\}$, a very popular method of interpolation is to use a spline function. Spline functions originate with the "Draftsman's spline", a thin wooden or plastic beam which could be flexed to pass through the points to be interpolated. A mathematical model for the Draftsman's spline is the cubic spline (Ahlberg, Nilson and Walsh (1967)), which is the unique function in $H = \{f(t), t \in T, f \in C^2$, and $f''(0) = f''(1) = 0\}$ that passes through the points $\{t_i, X(t_i)\}$ and minimizes the integral

$$\int_T [f''(t)]^2 dt \tag{1.1}$$

This integral criterion is referred to as the roughness of the function, and in this sense the spline is the smoothest function in H which passes through the points of interpolation.

The model of the Draftsman's spline is immediately generalized by letting L be a linear differential operator

$$L = \sum_{j=0}^{m} a_j D^j \tag{1.2}$$

where $D = \dfrac{d}{dt}$, $m > 0$ and $a_m = 1$. The so-called "L-spline of
interpolation" then is the unique function (Ahlberg, et al.
(1967)) in $H = \{f(t), t \in T, f^{(m-1)}(t)$ is absolutely continuous, $Lf \in L_2(T), f^{(j)}(0) = 0, j = 0, \ldots, m-1\}$ which passes
through the points of interpolation $\{t_i, X(t_i)\}$ and which
minimizes

$$\int_T [Lf(t)]^2 dt \tag{1.3}$$

A second related problem is that of smoothing splines.
A smoothing spline minimizes in $\hat{H}$

$$\int_T Lf(t)^2 dt + \sum_{i=1}^{n} \sum_{j=1}^{n} [f(t_i)-X(t_i)B^{ij}[f(t_j)-X(t_j)] \tag{1.4}$$

where B^{ij} is the i-j element of B^{-1} of a given positive
definite $n \times n$ matrix B. It can be shown that a "Smoothing
L-spline" is an interpolatory L-spline through a "smoothed"
set of ordinates (Kimeldorf and Wahba (1970)).

The computational formulas for the L-splines involve the
formal adjoint operator of L given by

$$L^* = \sum_{j=0}^{m} a_j (-D)^j \tag{1.5}$$

The L-splines then satisfy the $2m$-th order homogeneous
differential operator $L^*Lf(t) = 0$ for $t_i < t < t_{i+1}$, $i = 1, \ldots,$
$n-1$. Then by the conditions that $f(t_i) = X(t_i)$, $f(t) \in C^{2m-2}$
and the initial conditions $f^{(j)}(0) = 0$, there is sufficient
information to compute the parameters defining $f(t)$ in each
interval. Note also that the initial conditions $f^{(j)}(0) = 0$
are only for convenience as the integral conditions are invariant
to any solutions of $Lf^*(t) = 0$. Hence given any other set of
boundary conditions, let $f^*(t)$ be the solution to the
homogeneous equation which satisfies the boundary conditions

and let $\hat{f}(t)$ be the L-spline of interpolation to the points $X(t_i) - f^*(t_i)$; then the L-spline satisfying the desired boundary conditions exists and can be given by $f^*(t) + \hat{f}(t)$.

The Draftsman's spline (cubic spline) can be easily seen to satisfy this model with $L = D^2$. First $L^*Lf(t) = D^4 f(t) = 0$ in each interval implying a cubic polynomial (hence the name cubic spline). Also $f \in C^2$ where $2m - 2 = 2$ implies continuity of the first two derivatives.

An interesting generalization is to multivariate splines. Let

$$\underline{L} = \sum_{j=0}^{m} \underline{A}_j \, D^j \tag{1.6}$$

where $\underline{A}_j$ are $k \times k$ matrices, $\underline{A}_m = I$, and $\underline{L}\underset{\sim}{f}(t)$ $= \sum \underline{A}_j (D^j \underset{\sim}{f}(t))$. Also let $\underline{H}$ be the space of k dimensional vectors functions $\underset{\sim}{f}(t)$, $t \in T$, where $D^j \underset{\sim}{f}(0) = \underset{\sim}{0}$, $j = 0$, ..., $m-1$; $f_i^{(m-1)}(t)$ is absolutely continuous for $i = 1$, ..., k; and $\underline{L}\underset{\sim}{f}(t) \in L_2$, that is

$$\int_T [\underline{L}\underset{\sim}{f}(t)]' \, [\underline{L}\underset{\sim}{f}(t)] \, dt < \infty \tag{1.7}$$

<u>Definition 1.1.</u> Given the vectors $\{\underset{\sim}{Z}_1, \ldots, \underset{\sim}{Z}_n\}$, then the L-spline of interpolation is the unique function $\underset{\sim}{f}(t) \in \hat{\underline{H}}$ which passes through the vectors $\underset{\sim}{Z}_i$ and which minimizes

$$\int_T \underline{L}\underset{\sim}{f}(t)' \, \underline{L}\underset{\sim}{f}(t) \, dt \tag{1.8}$$

Note again that the initial conditions $D^j \underset{\sim}{f}(0) = \underset{\sim}{0}$ are only for convenience as they are easily changed both numerically and theoretically.

<u>Theorem 1.1</u> (Davis (1973)). Given $\{\underset{\sim}{Z}_1, \ldots, \underset{\sim}{Z}_n\}$, then the L-spline of interpolation is given by

$$\underset{\sim}{f}(t) = \sum_{j=1}^{n} \underline{K}(t,t_j)\, \underset{\sim}{a}_j \tag{1.9}$$

where

$$\underset{\sim}{Z}_i = \sum_{j=1}^{n} \underline{K}(t_i,t_j)\, \underset{\sim}{a}_j \tag{1.10}$$

and $\underline{K}(s,t)$ is the matrix Greens function of $\underline{L}^*\underline{L}$.

<u>Definition 1.2.</u> Given positive definite matrices $\{\underline{B}_1,\ \ldots,\ \underline{B}_n\}$, then the smoothing $\underline{L}$-spline is the unique function in $\hat{\underline{H}}$ which minimizes

$$\int_T \underline{L}\underset{\sim}{f}(t)'\,\underline{L}\underset{\sim}{f}(t)\,dt + \sum_{j=1}^{n} [\underset{\sim}{f}(t_j)-\underset{\sim}{Z}_j]'\,\underline{B}_j^{-1}\,[\underset{\sim}{f}(t_j)-\underset{\sim}{Z}_j] \tag{1.11}$$

<u>Corollary 1.1</u> (Davis (1973)). Given $\{\underset{\sim}{Z}_1,\ \ldots,\ \underset{\sim}{Z}_n\}$ and $\{\underline{B}_1,\ \ldots,\ \underline{B}_n\}$, the smoothing $\underline{L}$-spline is given by

$$\underset{\sim}{f}(t) = \sum_{j=1}^{n} \underline{K}(t,t_j)\, \underset{\sim}{a}_j \tag{1.12}$$

where

$$\underset{\sim}{Z}_i = \sum_{j=1}^{n} [\underline{K}(t_i,t_j) + \delta_{ij}\underline{B}_i]\, \underset{\sim}{a}_j \tag{1.13}$$

and

$$\delta_{ij} = 1 \text{ if } i = j, \text{ and } 0 \text{ if } i \neq j \tag{1.14}$$

<u>Corollary 1.2.</u> The smoothing $\underline{L}$-spline is an L-spline of interpolation through the "smoothed" vectors $\underset{\sim}{r}_j$, $j = 1,\ \ldots,\ n$, where

$$\underset{\sim}{v}_i = \sum_{j=1}^{n} \underline{K}(t_i,t_j) \; \underset{\sim}{r}_j \tag{1.15}$$

and

$$\underset{\sim}{z}_j = \sum_{i=1}^{n} [\underline{K}(t_j,t_i) + \delta_{ij} \; \underline{B}_i] \; \underset{\sim}{v}_i \tag{1.16}$$

2. CONTINUOUS TIME AUTOREGRESSIVE PROCESSES

Let $\underline{G}(s,t)$ be the matrix valued Greens function for $\underline{L}$ in $\hat{\underline{H}}$, and let $\underset{\sim}{V}_j(t)$, $j = 1, \ldots, m$ be a linearly independent basis for the null space of $\underline{L}$, then given the matrices $\underline{A}_1, \ldots, \underline{A}_n$ and the process $\underset{\sim}{W}(s)$, $s \in T$ with uncorrelated increments the linear process

$$\underset{\sim}{X}(t) = \sum_{j=1}^{m} \underline{A}_j \, \underset{\sim}{V}_j(t) + \int_T \underline{G}(t,s)\,d\underset{\sim}{W}(s) \tag{2.1}$$

is a model for continuous time autoregressive series. Such a process can be said to formally at least satisfy the differential equation $\underline{L}\underset{\sim}{X}(t) = d\underset{\sim}{W}(t)$, replacing the difference operator in discrete time with a differential operator in continuous time. The matrices $\underline{A}_j$ determine the initial conditions of the process which is considered fixed in this model. For a discussion in the univariate case of such processes with stochastic initial conditions, see Doob (1953), pg. 546.

The covariance kernel of the process is then given by

$$\underline{K}(s,t) = E\,\underset{\sim}{X}(s)\underset{\sim}{X}'(t) = \int_T \underline{G}(s,u)\underline{G}(t,u)\,du \tag{2.2}$$

It follows that

$$\underline{L}^*\underline{L}\underline{K}(s,t) = \int_T \underline{L}_s\underline{G}(s,u)\underline{L}_t\underline{G}(t,u)\,du = "\delta(t-s)" \tag{2.3}$$

that is that the covariance kernel of the process $\underset{\sim}{X}(t)$ is also the Green's function of $\underline{L}^{*}\underline{L}$ used in the definition of the L-splines. From Parzen (1967), it is known that the Best Linear Predictor of the process $\underset{\sim}{X}(t)$ is that linear combination

$$\hat{\underset{\sim}{X}}(t) = \sum_{j=1}^{n} \underline{c}_j(t)\, \underset{\sim}{X}(t_j) \tag{2.4}$$

such that $\underline{K}(t,t_i) = \sum_{j=1}^{m} \underline{c}_j(t)\, \underline{K}(t_i,t_j)$. From this and Theorem 1.1 follows:

Theorem 2.1. If for times $\{t_1, \ldots, t_n\}$ the values $\underset{\sim}{X}(t_i)$ are observed, then the Best Linear Predictor of the process $\underset{\sim}{X}(t)$ is the $\underline{L}$-spline with initial conditions given by $\sum \underline{A}_j \underset{\sim}{V}_j(0)$.

Similarly, one can prove:

Theorem 2.2. If one observes $\underset{\sim}{Y}_i = \underset{\sim}{X}(t_i) + \underset{\sim}{\varepsilon}_i$, $i = 1, \ldots, n,$ where the $\underset{\sim}{\varepsilon}$'s are serially uncorrelated observation error vectors with positive definite covariance matrices $\underline{B}_i$, then the Best Linear Predictor $\underset{\sim}{X}(t)$ is the smoothing $\underline{L}$-spline.

3. APPLICATION

The path of a point wandering in some k-dimensional space, which we will take to be a 2-dim plane, is represented by a vector indexed with time. Also there is often an orientation of the plane such that the vector process may well be represented as a vector of two uncorrelated auto-regressive processes, that is with covariance matrix

$$\underline{K}(s,t) = \begin{bmatrix} K_1(s,t) & 0 \\ 0 & K_2(s,t) \end{bmatrix} \tag{3.1}$$

It is then easily seen from the above Theorem 2 that the
solution to the Best Linear Predictor of the vector process
would have components that are the Best Linear Predictor of
each component process. However, as is often the case, the
observation error vectors will not be uncorrelated (for
example if the process is observed from a perspective oblique
to the plane).

If the process is observed with correlated observation
errors, Theorem 2.2 gives the Best Linear Predictor to be a
smoothing $\underline{L}$-spline. It follows from the two corollaries to
Theorem 1 that the smoothing $\underline{L}$-spline is an $\underline{L}$-spline of interpolation through a smoothed set of points, and that since the
matrix $\underline{K}$ is diagonal in this case, the interpolatory spline
will in this case be the vector of univariate interpolatory
L-splines. This result is of considerable computational importance as a direct solution will involve a kn by kn
matrix manipulation at every point $\hat{\underset{\sim}{X}}(t)$ is desired, which
the solution to the L-spline does not require.

To compute the smoothed vectors $\underset{\sim}{r}_j$, the following
iterative procedure can be shown to converge. Given the k-th
iterates $\underset{\sim}{r}_j^k$, j = 1, ..., n, let

$$\underset{\sim}{u}_j = \underline{B}_j^{-1} \ (\underset{\sim}{Y}_j - \underset{\sim}{r}_j^k) \tag{3.2}$$

then

$$\underset{\sim}{r}_i^{k+1} = \sum_{j=1}^{n} \underline{K}(t_i, t_j) \ \underset{\sim}{u}_j \tag{3.3}$$

Then once the smoothed vectors $\underset{\sim}{r}_j$ are computed, the Best
Linear Predictor will be the L-splines through the components
which satisfy the homogeneous equation L*Lf(t) = 0 subject
to the initial conditions and continuity conditions mentioned
before.

The spline most commonly used in the absence of any other
knowledge is the cubic spline. Interpolatory cubic spline
programs are in abundance. If a cubic spline is used, the
smoothed values $\underset{\sim}{r}_j$ should then be computed using for
$K_i(s,t)$ the function

$$
K_i(s,t) = \begin{cases} \dfrac{s^2 t}{2} - \dfrac{s^3}{3} & s \leq t \\[2ex] K_i(t,s) & s > t \end{cases}
\tag{3.4}
$$

BIBLIOGRAPHY

Ahlberg, J. H., E. N. Nilson and J. L. Walsh (1967).
    The Theory of Splines and Their Applications.
    New York: Academic Press.

Davis, H. T. (1973).   A Note on Multivariate L-Splines,
    Technical Report, University of New Mexico.

Doob, J. L. (1953). *Stochastic Processes*. New York:
    John Wiley & Sons.

Kimeldorf, G. and G. Wahba (1970). Spline functions
    and stochastic processes. *Sankhyā*, <u>132</u>,
    173-80.

Parzen, E. (1967). Statistical inference on time
    series by Hilbert space methods, I. *Time Series
    Analysis Papers*. San Francisco: Holden-Day.

PROBLEMS IN SELECTING A PATTERN RECOGNITION ALGORITHM

John W. Van Ness
The University of Texas at Dallas
Dallas, Texas

1. INTRODUCTION

Admissibility in the special adaption used for clustering
and discriminant analysis is introduced in Fisher and Van Ness
(1971 and 1973) and Van Ness (1973). Related discussions can
be found in Rubin (1967) and Jardine and Sibson (1968a). Gower
(1971) comments on the admissibility versus the axiomatic approach.
Classical admissibility in a discriminant analysis context is
also discussed for the nearest k-neighbor algorithms in Cover
and Hart (1967) and for the Gaussian case in Anderson (1958)
and elsewhere. Wright (1973) gives a set of axioms which determine a broad class of clustering functions.

If one attempts the axiomatic approach he usually runs
into the following difficulties:

1. in order to get uniqueness he must require conditions
   which at best are subject to question considering the
   known information; and

2. to make the mathematics tractable in solving for the
   algorithm, he must choose axioms which can lead only
   to the mathematically simplest algorithms such as
   nearest neighbor (e.g., see Johnson (1967) and
   Jardine and Sibson (1968b)) even though it is clear
   that other algorithms have merit.

On the other hand, the admissibility method adopts a differ- ent philosophy. A set of properties (admissibility conditions) are defined which might be desirable in some practical sit- uation. An algorithm satisfying, say, property A is called A-admissible. Hopefully a table (an enlargement of Table 1) would be available which gives an extensive list of admissibil- ity conditions across the top and a similarly extensive list of algorithms down the side. The user decides which of the ad- missibility conditions are important in his application and then looks for those algorithms which are admissible in all the senses he has chosen. He may find none, one or many. If he finds just a few he would probably try them all. If he finds many he at least will know he has eliminated some of the ob- viously bad algorithms.

This approach is a formalization of what, in effect, many people have been doing on a limited basis for some time. The challenge is in defining admissibility conditions which are intuitively appealing, mathematically tractable and discrim- inating. Such conditions provide insight into the practical differences among algorithms. There are, of course, many con- siderations which are relevant to the choice of algorithms but which are not normally phrased as admissibility conditions (though some could be). These include such things as required computer time, required computer memory, required data format, computational stability, etc. There are comments on this later in the paper.

2. THE DISCRIMINANT ANALYSIS PROBLEM

Discriminant analysis covers a class of problems of such diversity that people in different fields frequently view the problem from entirely different viewpoints and consider com- pletely different aspects of the problem to be important. Thus if one reads an article, say, on character recognition and another article on computer-assisted medical diagnosis, he

probably finds that even the discriminant analysis contents of
the two papers appear unrelated. Some have pointed out that
since the search for regularities in the environment is the
principle concern of scientific inquiry, the field of pattern
recognition could be considered as concerning the bulk of all
scientific endeavor. Also, the parallel development of related
methods in different fields has caused considerable terminolog-
ical and some conceptual confusion. Therefore, we try to define those aspects of the problem with which this paper is
concerned.

In the terminology as used, for example, in Kanal and
Chandrasekaran (1972), we first distinguish between the
linguistic model and the *statistical model* or correspondingly,
the *structural approach* and the *geometric approach*. [The Kanal
and Chandrasekaran article provides a good discussion of this.]
The linguistic approach roughly assumes that all possible
patterns in question are each comprised of a distinct combination of elements from a common class of *primitives*. Thus, in
case the patterns are alphabetic characters, the primitives
might include *vertical stroke*, *high loop*, etc. The linguistic
aspects of a pattern recognition problem are concerned with the
logic involved in the juxtaposition of primitives to form a
legitimate pattern. Of course, this concern for the generative
grammer must be highly specialized to the particular problem.

The recognition of the primitives is, itself, a discriminant analysis problem (usually with a lower level of complexity)
which frequently involves the statistical approach. It may be
purely statistical or a mixture. Thus, many of the better solutions to discriminant problems involve a combination or mixture
of both the linguistic and the statistical approaches. See
Grenander (1969a, 1969b).

This paper is concerned only with the purely statistical
or geometric aspects of a classification problem. We consider

statistical discriminant analysis (supervised pattern recognition) in its most common form. There is a population Ω
of elements which, by nature, is partitioned into disjoint
subpopulations (or clusters or classes or species or taxa)
$\Omega_1, \ldots, \Omega_M$. We can potentially observe p real-valued variables $X_1, \ldots, X_p$ for any point in Ω. Thus for each point
ω sampled, we have an observation vector $\underset{\sim}{x} = \underset{\sim}{x}(\omega) = (X_1(\omega),$
$\ldots, X_p(\omega))$. A distance function $d(\cdot,\cdot)$ is defined over all
possible pairs of observation vectors. Hopefully, $d(\underset{\sim}{x},\underset{\sim}{y})$
tends to be small if $\underset{\sim}{x}$ and $\underset{\sim}{y}$ are observations from elements
of the same class and larger if $\underset{\sim}{x}$ and $\underset{\sim}{y}$ are from different
classes. Sometimes d is ordinary Euclidean distance. Usually
we think of somehow randomly sampling from Ω so that $(X_1,$
$\ldots, X_p)$ is a vector of random variables.

We are given a set of n correctly classified samples
$S = \{(\underset{\sim}{x}_1, j_1), \ldots, (\underset{\sim}{x}_n, j_n)\}$ where $\underset{\sim}{x}_k = \underset{\sim}{X}(\omega_k)$ and $\omega_k \in \Omega_{j_k}$;
$k = 1, \ldots, n$. Define the set $T = \{\underset{\sim}{x}_1, \ldots, \underset{\sim}{x}_n\}$ to be the
observation vectors from this labeled data or training set, S.
The problem is to construct a decision rule d_S, which depends
only on S and perhaps known prior information, that makes
the fewest errors in classifying incoming unlabeled data
vectors. Thus for a point ω_0 whose classification is unknown,
we can observe only $\underset{\sim}{y} = \underset{\sim}{X}(\omega_0)$, and in our notation $d_S(\underset{\sim}{y}) = j$
means that d_S has decided that $\underset{\sim}{y}$ is an observation from an
element of Ω_j . This may or may not be correct.

In some situations S is obtained from a random sample
from Ω and in others it is from a deterministic sample.
Also, in some applications S may change with time: the
number of variables may change and new training data may come
in. This leads to sequential procedures and learning procedures (e.g., see Fu (1968)). Such procedures are of great
importance and are mentioned later; however, the algorithms and
admissibility conditions in Table 1 are for S fixed.

3. TYPICAL ALGORITHMS

There is a vast literature on discriminant analysis which
is spread over many disciplines including, for example,
statistics, computer science, business, medicine, biology,
engineering, psychology, mathematics, etc. In this literature
there are numerous algorithms proposed--many quite *ad hoc*.
The algorithms entered in Table 1 and defined below are typical.
An example is given in Section 5.

Let $T_j = \{\underset{\sim}{x}_k : \underset{\sim}{x}_k \in T$ and $\underset{\sim}{x}_k = \underset{\sim}{X}(\omega), \omega \in \Omega_j\}$ be all
the labeled observations coming from Ω_j . We assume
$T_j \neq \phi$, $j = 1, \ldots, M,$ unless otherwise specified. Let
$\underset{\sim}{y}$ denote an unclassified point.

<u>Definition 3.1. Nearest k-neighbors (k-NN)</u>. Let $V(\underset{\sim}{y})$ be the
k points in T which are closest to $\underset{\sim}{y}$. Classify $\underset{\sim}{y}$ into
that class having greatest representation in $V(\underset{\sim}{y})$. See,
e.g., Cover and Hart (1967). Ties can be handled in a variety
of ways and will not concern us here. The 1-NN rule is the
ordinary nearest-neighbor algorithm commonly found in the
literature. Usually we require n to be much larger than k.
Some users have found slight changes in k to cause uncom-
fortably large changes in the classification of test data.

<u>Definition 3.2. Farthest neighbor</u>. Let $\underset{\sim}{z}_j$ be the farthest
point in T_j from $\underset{\sim}{y}$ for $j = 1, \ldots, M.$ Classify $\underset{\sim}{y}$ into
that class whose $\underset{\sim}{z}$-point is closest to $\underset{\sim}{y}$. This rule tends
to work well when the observation in each class tend to lie in
spheres of equal size (e.g., see Kuiper and Fisher (1971)).

<u>Definition 3.3. Centroid</u>. Let $\underset{\sim}{z}_j$ be the centroid of the
observations in T_j, i.e., $\underset{\sim}{z}_j = \sum_{\underset{\sim}{x} \in T_j} \underset{\sim}{x} / |T_j|$ where $|T_j|$
is the cardinality of T_j. Classify $\underset{\sim}{y}$ into that class whose
sample centroid is closest. Here we must assume that the ob-

servations are in a linear space and that d is defined over this linear space so that it makes sense to discuss $d(\underset{\sim}{y},\underset{\sim}{z}_j)$.

<u>Definition 3.4. Average linkage</u>. This algorithm calculates the average distance between $\underset{\sim}{y}$ and the points in T_j (i.e., $d_j = \sum_{\underset{\sim}{x} \in T_j} d(\underset{\sim}{x},\underset{\sim}{y})/|T_j|$) and classifies $\underset{\sim}{y}$ into that class corresponding to the smallest average distance. This appears close to the centroid method but a difference occurs in Table 1. In clustering it has been found that average linkage usually behaves better than centroid (e.g., see Kuiper and Fisher (1971)).

<u>Definition 3.5. Least squares</u>. If the unclassified data arrives in a batch Y, the least squares algorithm treats the whole batch at once. It could be easily modified to treat each point individually but then it would be quite similar to centroid discrimination. The object is to partition Y into sets $Y_1, \ldots, Y_m$, with each point in Y_j being classified in the jth class $j = 1, \ldots, M$. We choose that partition which minimizes the loss function

$$L(Y_1, \ldots, Y_m) = \sum_{j=1}^{M} \sum_{\underset{\sim}{x} \in T_j \cup Y_j} d^2(\underset{\sim}{x},\underset{\sim}{z}_j) \qquad (3.1)$$

where

$$\underset{\sim}{z}_j = \sum_{\underset{\sim}{x} \in T_j \cup Y_j} \underset{\sim}{x}/|T_j \cup Y_j| \qquad (3.2)$$

In words, this procedure minimizes the sample total within cluster sum of squares where we force T_j into the jth cluster. This is an optimization procedure (i.e., one optimizes L over all partitions of Y) which implies that if Y has more than just a few points, it is not practical to attempt this even on a computer due to the incredible number of possible partitions

(e.g., see Fortier and Solomon (1966) and Fisher and Van Ness
(1971)). Various iterative techniques (e.g., see Friedman and
Rubin (1967)) which are quite adaptable to computers can be
used but they may give a local optimum.

Definition 3.6. Linear discriminant analysis. This is the most
common form of discriminant analysis and is not defined here
(e.g., see Anderson (1958), Ch. 6). Table 1 lists two cases
(for M=2): one where the true prior probabilities are known,
$(\frac{1}{2},\frac{1}{2})$, and one where they are estimated from S, where S
is obtained from a random sample in the same manner in which
the test data are selected.

Definition 3.7. Quadratic discriminant analysis. Again, this
is a classical procedure and can be found in Anderson (1958),
Ch. 6. We also consider the two cases described in algorithm
3.6.

Definition 3.8. Bayes' theorem. Procedures based on Bayes'
theorem are very important, have nice properties and are widely
used in practice. However, difficulties may develop as discussed below. These procedures assume that the observations
in T are a random sample usually in one of two senses:
(1) each observation is independent with first the class being
chosen according to some prior distribution $\pi_1, \ldots, \pi_M$ and
then the observation being taken following the conditional
distribution $f(x|\Omega_j)$ of the chosen class; or (2) the class
is chosen in some nonrandom fashion with the observation still
chosen according to $f(x|\Omega_j)$. In the nonparametric case with
absolutely continuous distributions, the $f(x|\Omega_j)$, j = 1, ..., M,
are usually estimated using a windowed or smoothed estimate
(Parzen (1962)) of the form

$$\hat{f}(x|\Omega_j) = \int w(x-y)\ dF_n(y|\Omega_j) \tag{3.3}$$

where $F_n(\cdot|\Omega_j)$ is the sample cumulative distribution function
for the data in T_j. This has inherent difficulties as discussed
in Section 7 of this paper. The window w has integral one
and is usually nonnegative, symmetric about the origin and de-
creasing with distance from the origin (see Parzen (1962) and
Van Ryzin (1969) for more information). For example, one
might use a multivariate normal density function for w. The
priors can either be known somehow or if sampling method (1)
above is used, they can be estimated from S. At this point
loss functions can be introduced but for ease of exposition
we assume the losses are 1 for misclassification and 0 for
correct classification.

Given the true or estimated priors $\pi_1, \ldots, \pi_2$
Bayes' theorem now gives us $\hat{f}(\Omega_j|\underset{\sim}{y})$, and we classify $\underset{\sim}{y}$ into
that class having highest estimated *a posteriori* probability.
In Table 1 we assume the true priors are known and equal.

In the discrete distribution case, the obvious estimates
of $f(\underset{\sim}{x}|\Omega_1)$ and $f(\underset{\sim}{x}|\Omega_2)$ are used.

4.   TABULATED ADMISSIBILITY CONDITIONS

The following admissibility conditions are defined in
Fisher and Van Ness (1973). Some are motivated by corresponding
conditions in clustering.

<u>Definition 4.1. Well-structured</u>. This is one of the most
appealing conditions. Suppose the data happens to come out in
such a way that the classification appears obvious, then we
require the algorithm to make the obvious classification.
Specifically we say the data is well-structured if there is a
partition $Y_1, \ldots, Y_M$ of Y such that the M sets
$Y_1 \cup T_1, \ldots, Y_M \cup T_M$ have the property that all within-set
interpoint distances are smaller than all between-set distances,
i.e.

$$d(\underset{\sim}{x},\underset{\sim}{y}) < d(\underset{\sim}{u},\underset{\sim}{v}) \tag{4.1}$$

for all x,y in the same set and all u,v in different sets.
An algorithm is well-structured admissible if it classifies
all $y \in Y_j$ into class j, $j = 1, \ldots, M$, whenever the data
is well-structured (see Rubin (1967)).

Definition 4.2. __Monotone__. If it is impossible to assign a
scale to the variables (see Johnson (1967)), it would be nice
for the algorithm to be scale invariant. Let h be a continuous monotone increasing function with $h(0) = 0$. Define
a new distance function $d'(x,y) = h(d(x,y))$, $x \in T \cup Y$,
$y \in T \cup Y$. If an algorithm discriminates in the same manner,
using d and d' for any T, Y and h, it is called
monotone admissible.

Definition 4.3. __Convex__. Data are said to be convex if the
respective convex hulls $H_1, \ldots, H_M$ of $T_1, \ldots, T_M$ are
disjoint and if there is a partition $Y_1, \ldots, Y_M$ of Y such
that $Y_j \subset H_j$, $j = 1, \ldots, M$. The data must lie in a linear
space for the convex hulls to be defined. An algorithm is convex admissible if it classifies all $y \in Y_j$ into class j
for $j = 1, \ldots, M$ whenever it sees convex data.

Definition 4.4. __Random Convex__. The condition 4.3 is too strong,
i.e., none of our algorithms are convex admissible. This is
partially due to the fact that we are allowed to choose the convex data deterministically and pathologically. Therefore,
we made the following modification. Suppose $H_1, \ldots, H_M$
are disjoint as in 4.3 and we put a uniform probability distribution P_j over H_j. We then sample y according to P_j
and require that the probability that y is classified by the
algorithm into class j is at least as great as the probability that y is classified into any other class. This must
be true for all j and all T with $H_1, \ldots, H_M$ disjoint.
Definition 4.5. __Repeatable__. This condition applies only to
those algorithms which have corresponding clustering algorithms.
This is true of the first five algorithms defined in Section 3

(e.g., see Fisher and Van Ness (1971) for the corresponding clustering procedures). If a clustering $C_1, \ldots, C_M$ arises from one such clustering algorithm it might be considered as "natural" in that sense. Suppose $C_1, \ldots, C_M$ is such a clustering. We remove an arbitrary point $\underset{\sim}{y}$ from $C_1 \cup \cdots \cup C_M$, say $\underset{\sim}{y} \in C_k$, and take $T_j = C_j$ for $j \neq k$ and $T_k = C_k - \{\underset{\sim}{y}\}$. We then apply the corresponding discriminant algorithm to $\underset{\sim}{y}$ and say it is repeatable admissible if $\underset{\sim}{y}$ is reclassified into C_k for any clustering $C_1, \ldots, C_M$ so obtained.

Definition 4.6. <u>Equal</u>. If the conditional distributions $f(\cdot | \Omega_1), \ldots, f(\cdot | \Omega_M)$ were all identical, then the data would give no information with which to discriminate the classes. If, in addition, $\pi_1 = \ldots = \pi_M = 1/M$, then the maximum probability of correctly classifying a randomly chosen point $\underset{\sim}{y}$ is $1/M$. However, if the conditionals are not all equal, then there is information in the data and we can do better conceptually than $1/M$. An algorithm is called equal admissible if for $\pi_1 = \ldots = \pi_M$ and conditional distributions which are mixtures of discrete and absolutely continuous parts, the probability of correctly classifying a randomly chosen point is $\geq 1/M$.

Definition 4.7. <u>Asymptotically equal</u>. As is noted in Section 6, equal admissibility is too strong a requirement. Therefore, we modify the above definition to require that the limit of the probability of correctly classifying a randomly chosen $\underset{\sim}{y}$ is $\geq 1/M$ as the number of points in T tends to infinity.

Table 1 contains tabulations for the algorithms and admissibility conditions just defined. For proof of the entries see Fisher and Van Ness (1973).

5. AN EXAMPLE

For the uninitiated we give an example which illustrates many of the concepts of Sections 3 and 4. Let $M=3$, $n=10$,

TABLE 1

ADMISSIBILITY TABLE

Admissibility Condition / Discriminant Algorithm	Well-structured	Convex	Random convex	Monotone	Equal	Asymptotic equal	Repeatable
Nearest neighbor	yes	no	no	yes	no	yes	yes
Farthest neighbor	yes	no	no	yes	no	no	no
Centroid	no	no	no	no	no	no	no
Average linkage	yes	no	no	no	no	no	no
Least squares	no	no	no	no	no	no	yes
Linear ($\frac{1}{2},\frac{1}{2}$)	no	no	no	no	no	no	
Linear (est.)	no	no	no	no	no	no	
Quadratic ($\frac{1}{2},\frac{1}{2}$)	no	no	no	no	no	no	
Quadratic (est.)	no	no	no	no	no	no	
Bayes' theorem	yes	no	no	no	no	yes	

$p=2$, d be Euclidean and the data be arranged as shown in Figure 1. Here $T_1 = \{a,b,c,d\}$, $T_2 = \{e,f,g\}$, $T_3 = \{h,i,j\}$ and $Y = \{z\}$.

Using the nearest-neighbor algorithm, z would be in class 1 since d is closest to z. Farthest-neighbor would classify z in class 3 since a and g are both farther from z than h. Centroid would classify z as a 3 since the centroid of T_3, (6.67, 5), is closer to z than those

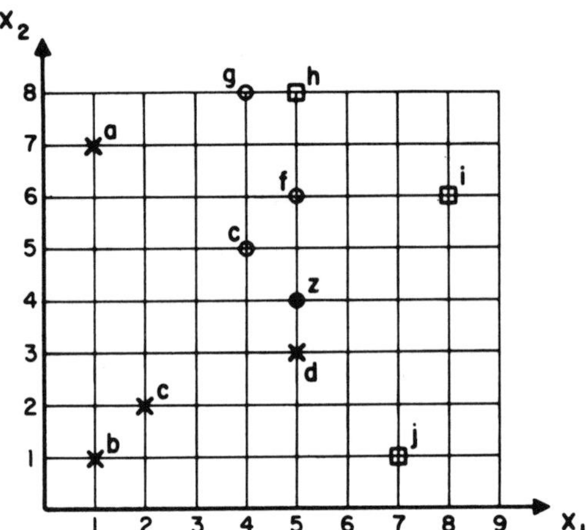

Figure 1. An example of data.

of T_1 and T_2. Average linkage would classify z as a 2
since d_2 = 7.53/5 = 2.51 which is smaller than d_1 or d_3.
Least squares would also classify z as a 2 since
$L(\phi,\{z\},\phi)$ = 6.5 which is much smaller than L for any other
classification. The results of the Bayes algorithm would depend on the window w used. Specht (1967) has some diagrams
which illustrate the Bayes algorithm.

This data obviously is not well structured since, for
example, the within-cluster distance between a and d is
much larger than the between-cluster distance between e and
d. Thus various well-structured admissible rules may classify
z differently for this data. If one plots some well-structured
data it is seen that they are obviously grouped as they should
be.

The three convex hulls H_1, H_2 and H_3 are disjoint for
this data but z does not fall within any of these hulls.
So, again, various convex admissible rules may treat z differently.

6. RANDOM ADMISSIBILITY CONDITIONS

As mentioned in Fisher and Van Ness (1973), the listed
admissibility conditions are too strong, thus resulting in
the preponderance of no's in Table 1. The reason for this is
that an admissible algorithm must behave properly for all
sets of data satisfying the conditions, and this data may be
chosen deterministically (except partially for random convex
admissible). By choosing a pathological set of data one can
cause the algorithm to fail to satisfy a seemingly reasonable
admissibility condition. Except for, perhaps, the well-
structured, monotone and repeatable admissibilities the conditions are not, in fact, reasonable.

To avoid these difficulties we need "random admissibility
conditions". These conditions impose restrictions on the con-
ditional probabilities $f(\underset{\sim}{x}|\Omega_j)$, j = 1, ..., M, and then re-
quire the algorithm to behave correctly with high probability
for data generated according to

$$f(\underset{\sim}{x}) = \sum_{j=1}^{M} \pi_j \, f(\underset{\sim}{x}|C_j)$$

To do this one needs means of quantifying the "niceness" of a
probability measure. One approach to this is to let the
$f(\underset{\sim}{x}|\Omega_j)$'s be members of some common parametric family or, perhaps, simple mixtures of members of such a family. Some work
has been done on the behavior of some algorithms when data
are from particular families. Presently there is considerable
interest in calculating the probability of misclassification
under special conditions (e.g., see Hills (1966), Lachenbruch
and Mickey (1968) and Sorum (1973)). If this work can be
broadened, it can be conveniently phrased to provide some useful admissibility criteria. A second approach is to find
some tractable nonparametric measures of "niceness" for
distributions.

7. OTHER CONSIDERATIONS IN ALGORITHM SELECTION

In this section we take up some further considerations in
the selection of an algorithm not directly related to the admissibility approach. In particular, we wish to comment on
the commonly used (and sometimes misused) Bayes procedures and
on the perhaps too infrequently used sequential procedures.

Bayes procedures have certain optimal properties which
would argue for their use. In fact, they are admissible in the
classical sense. An excellent reference on their use is Fisher,
Kronmal and Diehr (1973).

In order to determine a Bayes decision rule one must use
the conditional densities $f(\cdot|\Omega_j)$, $j = 1, \ldots, M$, mentioned
in Section 3. If one can justify using a certain parametric
family, such as the Gaussian family, so that to train on the
training data one need only estimate a few (relative to the
given training sample size in each class) parameters, then
Bayes procedures are obvious candidates for usable algorithms.
However, if the problem is very nonparametric in nature or
the number of parameters is large relative to the training
sample size (as is frequently the case in medical problems,
for example), then great care must be used in dealing with
Bayes estimates. This latter situation is likely to be true
when the dimension p is large.

When the $f(\cdot|\Omega_j)$, $j = 1, \ldots, M$, are estimated and not
given, the procedure loses its Bayes optimality but in the parametric situation one still expects a good decision rule. The
problem comes when $f(\cdot|\Omega_j)$ cannot be well-estimated. Con-
sider, for example, the simple situation in which $p = 10$ and
each X_j, $j = 1, \ldots, 10$, is discrete with only four possible
values. To estimate $f(\cdot|\Omega_j)$ would require estimating over
one million probabilities (parameters). The continuous nonparametric situation is even more hopeless.

Several modifications have been suggested to overcome this
problem. We briefly mention a few here.

1) Assume the variables $X_1, \ldots, X_p$ are independent (within each class). This crude-sounding and often unrealistic assumption frequently leads to good results (see Gilbert (1968), Fisher, Kronmal and Diehr (1973), and Ledley and Lusted (1959)).

2) Assume a reasonably flexible parametric family even though it cannot be completely justified and revert to a parametric situation. This is what one does in effect when he uses linear or quadratic discriminant analysis.

3) Allow some sort of limited dependency between variables. For example, one could assume that there are certain groups of the variables which are independent from one another. This allows for a more realistic model at the cost of increased estimation difficulties. Lincoln and Parker (1967) have a novel idea along these lines: they assume that $X_1, \ldots, X_p$ form a Markov process. This also allows for a limited dependency among the variables.

4) Reduce the dimension of the problem by using some method which fits the fact that the goal is to discriminate and not simply to get a set of "orthogonal variables." One of the best methods is simply to ask for help from experts in the field being studied. Various "feature selection" procedures are in the literature (e.g., see Patrick (1972)).

Many of the above procedures destroy the original optimality properties claimed for a Bayes procedure, even in an asymptotic sense; thus it is unlikely that they would be admissible in nearly as many senses as would a genuine Bayes procedure. One sees that care must be taken in using any future admissibility table.

Another consideration in the use of Bayes procedures for classifying large amounts of data using large training data sets is, of course, memory size and computational feasibility. A straightforward approach may require having in memory all training data to classify each point and may also have no way of using the calculations made to classify any one point to reduce the calculations required to classify later points.

Specht (1967) proposes a method by which a polynomial expansion
of a classification rule, based on Parzen-type density estimate
using Gaussian windows, is used to allow for an up-dating procedure for calculating the discriminant functions and for the
storage of a few coefficients instead of the training data.
This method has been criticized as being too sensitive to out-
liers.

Finally, this author would like to advocate to anyone
about to select an algorithm, that he keep in mind sequential
procedures. Many classification problems naturally proceed
sequentially--this is true, for example, in medical diagnosis.
Decision trees are very useful in many applications. Often
considerable savings can be accrued by using sequential methods
(see Garry and Barnett (1968)). Sequential methods of great
variety have been proposed--some using, for example, information theory, dynamic programming, classical statistical
methods, etc. A good review of the area can be found in Fisher,
Kronmal and Diehr (1973).

ACKNOWLEDGMENT

Research for this paper was supported by NSF Grant
GP-37984 at Carnegie-Mellon University.

BIBLIOGRAPHY

Anderson, T. W.   (1958). *Introduction to Multivariate
Statistical Analysis*. New York: John Wiley and
Sons, Inc.

Cover, T. M. and Hart, P. E.   (1967). Nearest neighbor
pattern classification. *IEEE Transactions on Information Theory* IT-13, 21-7.

Fisher, L., Kronmal, R. and Diehr, P.   (1973).
"Mathematical aids to medical decision making,"
manuscript, Dept. of Biostatistics, Univ. of
Washington, Seattle.

Fisher, L. and Van Ness, J.   (1971).   Admissible
    clustering procedures. *Biometrika* 58 (1), 91-104.

Fisher, L. and Van Ness, J.   (1973).   Admissible
    discriminant analysis. *J. Amer. Statist. Assoc.*
    68 (343), 603-7.

Fortier, J. J. and Solomon, H.   (1966).   Clustering
    procedures. *Multivariate Analysis* (ed., P. R.
    Krishnaiah). New York: Academic Press, 493-506.

Friedman, H. P. and Rubin, J. (1967).   On source invariant
    criteria for grouping data.   *J. Amer. Statist. Assoc.*
    62, 1159-78.

Fu, K. S. (1968).   *Sequential Methods in Pattern Recognition
    and Machine Learning*.   New York: Academic Press.

Garry, G. A. and Barnett, G. O. (1968).   Experience with a
    model of sequential diagnosis. *Comp. Biomed. Res.* 1,
    470-507.

Gilbert, E. S. (1968).   On discrimination using qualitative
    variables.   *J. Amer. Statist. Assoc.* 63, 1399.

Gower, J. C. (1971).   "Classification criteria and algorithms",
    unpublished manuscript.

Grenander, U. (1969).   Foundations of pattern analysis.
    Quart. Appl. Math. XXVII (1), 1-55.

Grenander, U. (1969).   "A unified approach to pattern analy-
    sis".   Tech. Report, Brown Univ. Providence, R.I.

Hills, M. (1966).   Allocation rules and their error rates.
    J. R. Statist. Soc. (Series B) 28, 1-32.

Jardine, N. and Sibson, R. (1968).   The construction of h
    hierarchic and nonhierarchic classifications.
    Computer J. 11, 177-84.

Jardine, N. and Sibson, R. (1968).   A model for taxonomy.
    Math. Biosci. 2, 465-82.

Johnson, S. C. (1967). Hierarchical clustering schemes.
    Psychometrika <u>32</u>, 241-5.

Kanal, L. and Chandrasekaran, B. (1972). On linguistic,
    statistical and mixed models for pattern recognition.
    Frontiers of Pattern Recognition (S. Watanabe, Ed.).
    New York: Academic Press, 163.

Kuiper, K. and Fisher, L. (1971). "A Monte-Carlo com-
    parison of six clustering procedures." M.S. thesis,
    University of Washington, Department of Mathematics.

Lachenbruch, P. A. and Mickey, M. R. (1968). Estimation
    of error rates in discriminant analysis. *Techno-
    metrics* <u>10</u>, 1-11.

Ledley, R. and Lusted, L. (1959). The reasoning foundations
    of medical diagnosis. *Science* <u>130</u>, 9.

Lincoln, T. and Parker, R. (1967). Medical diagnosis using
    Bayes' theorem. *Health Serv. Res.*, 34-45.

Patrick, E. A. (1972). *Fundamentals of Pattern Recognition*.
    New Jersey: Prentice-Hall.

Parzen, E. (1962). On estimation of a probability density
    function and mode. *Ann. Math. Statist.* <u>33</u> (3), 1065-76.

Rubin, J. (1967). Optimal classification into groups: an
    approach for solving the taxonomy problem. *J. Theoret.
    Biol.* <u>15</u>, 103-44.

Sorum, M. (1973). Estimating the expected probability of
    misclassification for a rule based on the linear dis-
    criminant function: univariate normal case.
    Technometrics <u>15</u>, 329-39.

Specht, D. F. (1967). Generation of polynomial discriminant
    functions of pattern recognition. *IEEE Transactions
    on Electronic Computers* <u>EC-16</u> (3), 308-19.

Van Ness, J. (1973). Admissible clustering procedures.
    Biometrika <u>60</u> (2), 422.

Van Ryzin, J. (1969). On strong consistency of density
    estimates. *Ann. Math. Statist.* <u>40</u>, 5.

Wright, W. E. (1973). A formalization of cluster analysis.
    Pattern Recognition <u>5</u> (3), 273-82.

ON MINIMIZING THE PROBABILITY OF
MISCLASSIFICATION FOR LINEAR FEATURE SELECTION:
A CLASSIFICATION PROCEDURE

L. F. Guseman, Jr.

Texas A&M University
College Station, Texas

Homer F. Walker

University of Denver
Denver, Colorado

1. INTRODUCTION

Consider two populations Π_1 and Π_2 with associated multivariate normal density functions defined for $x = (x_1, \ldots, x_n)^T \in E^n$ by

$$p_i(x) = (2\pi)^{-n/2} |\Sigma_i|^{-1/2} \exp[-\frac{1}{2}(x-\mu_i)^T \Sigma_i^{-1}(x-\mu_i)], \quad 1 = 1, 2$$

(1.1)

where μ_1, μ_2, Σ_1, and Σ_2 are known parameters. If $B = (b_1, \ldots, b_n)$ is a nonzero $1\times n$ vector and $x \in E^n$, then $Bx \in E^1$ and the populations Π_1 and Π_2 have transformed normal density functions defined for $y \in E^1$ by

$$p_i(y,B) = (2\pi)^{-1/2}(B\Sigma_i B^T)^{-1/2} \exp\left(-\frac{(y-B\mu_i)^2}{2B\Sigma_i B^T}\right), \quad i = 1, 2$$

(1.2)

Using a Bayes optimal (maximum likelihood) classification scheme, and assuming that *a priori* probabilities are equal that an observation comes from either Π_1 or Π_2, the transformed probability of misclassification in E^1 as a function of B, denoted by g, is given in Anderson (1958) by

61

$$g(B) = \frac{1}{2} \int_{R_1(B)} p_2(y,B)\,dy + \frac{1}{2} \int_{R_2(B)} p_1(y,B)\,dy \tag{1.3}$$

where

$$R_1(B) = \{y \in E^1 : p_1(y,B) \geq p_2(y,B)\} \tag{1.4}$$

and

$$R_2(B) = \{y \in E^1 : p_1(y,B) < p_2(y,B)\} \tag{1.5}$$

(If $p_1(y,B) \equiv p_2(y,B)$, we define $g(B) = \frac{1}{2}$.) Letting

$$L(y,B) = \log \frac{p_1(y,B)}{p_2(y,B)}$$

$$= \frac{1}{2} \log \frac{B\Sigma_2 B^T}{B\Sigma_1 B^T} - \frac{(y-B\mu_1)^2}{B\Sigma_1 B^T} + \frac{(y-B\mu_2)^2}{2B\Sigma_2 B^T} \tag{1.6}$$

and setting $L(y,B) = 0$, we obtain the equation

$$F(y,B) = \alpha(B)y^2 + 2\beta(B)y + \gamma(B) = 0 \tag{1.7}$$

where

$$\alpha(B) = B(\Sigma_1 - \Sigma_2)B^T \tag{1.8}$$

$$\beta(B) = -B(\Sigma_1 B^T B\mu_2 - \Sigma_2 B^T B\mu_1) \tag{1.9}$$

and

$$\gamma(B) = B\Sigma_1 B^T (B\mu_2)^2 - B\Sigma_2 B^T (B\mu_1)^2 + (B\Sigma_1 B^T)(B\Sigma_2 B^T)\log \frac{B\Sigma_2 B^T}{B\Sigma_1 B^T}$$

$$\tag{1.10}$$

Then, in terms of F,

$$R_1(B) = \{y \in E^1 : F(y,B) \geq 0\} \tag{1.11}$$

and

$$R_2(B) = \{y \in E^1 : F(y,B) < 0\} \tag{1.12}$$

Then the resulting problem is to determine a nonzero $1 \times n$
vector B which minimizes g. If g is to be minimized as a
function of B, it is natural to ask whether g is a differentiable function of the elements of B. If such is the
case, then the minimum value of g will occur only if these
derivatives all vanish.

In Section 3 we present a computational procedure for
determining a nonzero $1 \times n$ vector which minimizes g. The
procedure is based on theoretical results presented in Guseman
and Walker (1973a). Statements of these results are given in
Section 2 for the sake of completeness. In Section 4 we
present some preliminary numerical results. Concluding remarks concerning current research efforts are presented in
Section 5.

We have found it convenient to work with the *Gateaux
differential of* g *at* B *with increment* C, denoted by
$\delta g(B;C)$, and defined (if the limit exits) by

$$\delta g(B;C) = \lim_{s \to 0} \frac{g(B+sC)-g(B)}{s}$$

for a $1 \times n$ vector C. If for a given B the above limit
exists for each $1 \times n$ vector C, then g is said to be
Gateaux differentiable at B (Luenberger (1969), p. 171).
If g is *Gateaux differentiable* at B, then the derivative
of g with respect to, say, the jth component of B is given
by $\delta g(B;C_j)$, where C_j is the $1 \times n$ vector with a 1 in
the jth slot and zeros elsewhere. Similarly, if B is a nonzero $1 \times n$ vector, and C is a $1 \times n$ vector, we define

$$\delta p_i(y,B;C) = \lim_{s \to 0} \frac{p_i(y,B+sC) - p_i(y,B)}{s} , \quad i = 1, 2$$

To conclude this introduction, we observe that if a B
which minimizes g has been found, then the regions $R_1(B)$
and $R_2(B)$ defined by B are used for classification purposes

as follows: Given an observed vector x in E^n, compute $y = Bx$, and classify x as coming from Π_i if y is in $R_i(B)$, $i = 1, 2$. It is evident that the regions determined by the minimizing B give rise to a classification procedure which is different in general from that presented in Anderson and Bahadur (1962). This is a consequence of the fact that the discriminant function F for the regions $R_1(B)$ and $R_2(B)$ is quadratic, except in the special case $B\Sigma_1 B^T = B\Sigma_2 B^T$, even if $\Sigma_1 \neq \Sigma_2$.

2. DIFFERENTIATING THE PROBABILITY OF MISCLASSIFICATION

In this section we state results from Guseman and Walker (1973a) which show that, for a nonzero $1{\times}n$ vector B, $\delta g(B;C)$ exists for each $1{\times}n$ vector C. We also obtain a formula (Theorem 2.2) for $\delta g(B;C)$ which is numerically tractable. For the discussion of these results, we let

$$S(B) = \{y \in E^1 : F(y,B) = 0\} \tag{2.1}$$

As was shown in Guseman and Walker (1973a), if $p_1(y,B) \neq p_2(y,B)$, then $F(y,B)$ is a quadratic function of y, and $S(B)$ consists of at most two points; that is, $S(B) = \{a\}$, or $S(B) = \{a_1,a_2\}$ for some a, a_1, a_2 in E^1 (with $a_1 < a_2$). In the statement of Theorem 2.2, we use the notation

$$f(y)\Big|_{S(B)} = \begin{cases} f(a), & \text{if } S(B) = \{a\} \\ f(a_2) - f(a_1), & \text{if } S(B) = \{a_1,a_2\} \end{cases} \tag{2.2}$$

for a function f on E^1.

<u>Theorem 2.1.</u> Let B be a nonzero $1{\times}n$ vector. Then $\delta g(B;C)$ exists for each $1{\times}n$ vector C and is given by

$$\delta g(B;C) = \begin{cases} 0, & \text{if } p_1(y,B) \equiv p_2(y,B) \\[2ex] \dfrac{1}{2}\displaystyle\int_{R_1(B)} \delta p_2(y,B;C)\,dy + \dfrac{1}{2}\displaystyle\int_{R_2(B)} \delta p_1(y,B;C)\,dy \\[3ex] \hspace{6em} \text{if } p_1(y,B) \not\equiv p_2(y,B) \end{cases}$$

$$(2.3)$$

We remark that Theorem 2.1 remains true when B is a $k \times n$ matrix of rank k. A proof of this fact appears in Guseman and Walker (1973b). Unfortunately, when k is greater than 1, the formulas for the derivatives of g (as well as for g) do not appear to be numerically tractable.

According to Theorem 2.1, for nonzero B the Gateaux differential $\delta g(B;C)$ exists and can be calculated using the Gateaux differentials $\delta p_1(y,B;C)$ and $\delta p_2(y,B;C)$ of the transformed density functions. In the result below, we give the expression for $\delta p_i(y,B;C)$. For convenience, we omit subscripts.

<u>Lemma 2.1.</u> Let B be a nonzero $1 \times n$ vector. Then

$$\delta p(y,B;C) = -p(y,B) \left\{ \frac{C\Sigma B^T}{B\Sigma B^T} - \frac{C\mu}{B\Sigma B^T}(y-B\mu) - \frac{C\Sigma B^T}{(B\Sigma B^T)^2}(y-B\mu)^2 \right\}$$

$$(2.4)$$

for each $1 \times n$ vector C.

Theorem 2.1 and Lemma 2.1 provide an explicit formula for $\delta g(B;C)$. Unfortunately, this formula does not lend itself to easy computation because of the integrals that appear. Remarkably enough, a short calculation yields the formula of Theorem 2.2 below, in which no integrals appear.

<u>Theorem 2.2.</u> Let B be a nonzero $1 \times n$ vector. Then $\delta g(B;C)$ exists for each $1 \times n$ vector C and is given by

$$
\delta g(B;C) = \begin{cases} 0, \quad \text{if } p_1(y,B) \equiv p_2(y,B) \\[2ex] \pm \dfrac{1}{2}\, p_1(y,B)\left[C(\mu_2-\mu_1) + \dfrac{C\Sigma_2 B^T}{B\Sigma_2 B^T}\,(y-B\mu_2) - \dfrac{C\Sigma_1 B^T}{B\Sigma_1 B^T}\,(y-B\mu_1)\right]\Bigg|_{S(B)} \\[2ex] \quad \text{if } p_1(y,B) \not\equiv p_2(y,B), \text{ where the sign is} \\[1ex] \quad \text{taken to be} \\[1ex] \quad \lim_{y\to\infty}\,[\text{sign log}\,\{p_1(y,B)/p_2(y,B)\}] \end{cases}
$$

$$(2.5)$$

3. COMPUTATIONAL PROCEDURE

It is well known (Luenberger (1969), p. 178) that when B is an extremum of g, then $\delta g(B;C) = 0$ for each $1\times n$ vector C. It follows that if B minimizes g, then B must satisfy the vector equation

$$
\frac{\partial g}{\partial B} \equiv \begin{pmatrix} \delta g(B;C_1) \\ \cdot \\ \cdot \\ \cdot \\ \delta g(B;C_n) \end{pmatrix} = \begin{pmatrix} 0 \\ \cdot \\ \cdot \\ \cdot \\ 0 \end{pmatrix} \tag{3.1}
$$

where C_j, $1 \le j \le n$, is a $1\times n$ vector with a one in the jth slot and zeros elsewhere. Using the formula for $\dfrac{\partial g}{\partial B}$ provided by Theorem 2.2, one can employ existing minimization procedures based on this equation to find a local minimum of g.

Assuming that both the $n\times 1$ mean vectors μ_1 and μ_2 and the $n\times n$ covariance matrices Σ_1 and Σ_2 are known, we describe below a way in which the necessary functions can be computed for such a minimization procedure.

The error function

$$
\Phi(a) = (2\pi)^{-1/2} \int_{-\infty}^{a} \exp(-\frac{1}{2}\,t^2)\,dt \tag{3.2}
$$

can be computed using double precision library routines. Then, for a given nonzero $1\times n$ vector B, one can define

$$D_i(a,B) = \int_{-\infty}^{a} p_i(y,B)\,dy, \quad i = 1, 2 \tag{3.3}$$

and compute the values of $D_i(a,B)$ using the relationship

$$D_i(a,B) = \Phi\left(\frac{a - B\mu_i}{(B\Sigma_i B^T)^{1/2}}\right), \quad i = 1, 2 \tag{3.4}$$

After computing the scalars $B\mu_1$, $B\mu_2$, $B\Sigma_1 B^T$, and $B\Sigma_2 B^T$, one solves the quadratic equation

$$F(y,B) = \alpha(B)y^2 + 2\beta(B)y + \gamma(B) = 0 \tag{3.5}$$

where $\alpha(B)$, and $\beta(B)$ and $\gamma(B)$ are given by (1.8), (1.9), and (1.10). As noted in Section 2 this quadratic has either a single root or else two distinct real roots. These cases are treated separately in computing $g(B)$ and $\frac{\partial g}{\partial B}$.

3.1. Single Root Case

Equation (3.5) has a single root precisely when $\alpha(B) = 0$; that is, when the transformed covariances are equal. In this case, the single root, a, of (3.5) is given by $a = (B\mu_1 + B\mu_2)/2$. Then

$$g(B) = \begin{cases} \dfrac{1}{2} - \dfrac{1}{2}[D_1(a,B) - D_2(a,B)], & \text{if } R_1(B) = (-\infty, a] \\[2ex] \dfrac{1}{2} + \dfrac{1}{2}[D_1(a,B) - D_2(a,B)], & \text{if } R_2(B) = (-\infty, a) \end{cases} \tag{3.6}$$

and

$$\frac{\partial g}{\partial B} = \begin{cases} \mu_1 - \mu_2 - \dfrac{(\Sigma_1 + \Sigma_2)B^T}{B\Sigma_1 B^T + B\Sigma_2 B^T}(B\mu_1 - B\mu_2), & \text{if } R_1(B) = (-\infty, a] \\[3ex] \mu_2 - \mu_1 + \dfrac{(\Sigma_1 + \Sigma_2)B^T}{B\Sigma_1 B^T + B\Sigma_2 B^T}(B\mu_1 - B\mu_2), & \text{if } R_2(B) = (-\infty, a) \end{cases} \tag{3.7}$$

3.2. Two Root Case

Equation (3.5) has two distinct real roots if $\alpha(B) \neq 0$. In this case, let a_1, a_2 denote the roots of (3.5), arranged so that $a_1 < a_2$. Then,

$$g(B) = \begin{cases} \dfrac{1}{2} - K, & \text{if } R_1(B) = [a_1, a_2] \\[2em] \dfrac{1}{2} + K, & \text{if } R_2(B) = (a_1, a_2) \end{cases} \tag{3.8}$$

where

$$K = \frac{1}{2} \left\{ [D_1(a_2, B) - D_1(a_1, B)] - [D_2(a_2, B) - D_2(a_1, B)] \right\} \tag{3.9}$$

and

$$\frac{\partial g}{\partial B} = \begin{cases} K_1 \mu_1 - K_2 \mu_2 + K_3 \dfrac{\Sigma_1 B^T}{B \Sigma_1 B^T} - K_4 \dfrac{\Sigma_2 B^T}{B \Sigma_2 B^T} \,, & \text{if } R_1(B) = [a_1, a_2] \\[2.5em] K_2 \mu_2 - K_1 \mu_1 + K_4 \dfrac{\Sigma_2 B^T}{B \Sigma_2 B^T} - K_3 \dfrac{\Sigma_1 B^T}{B \Sigma_1 B^T} \,, & \text{if } R_2(B) = (a_1, a_2) \end{cases} \tag{3.10}$$

where

$$K_1 = P_1(a_2, B) - P_1(a_1, B) \tag{3.11}$$

$$K_2 = P_2(a_2, B) - P_2(a_1, B) \tag{3.12}$$

$$K_3 = (a_2 - B\mu_1)P_1(a_2, B) - (a_1 - B\mu_1)P_1(a_1, B) \tag{3.13}$$

$$K_4 = (a_2 - B\mu_2)P_2(a_2, B) - (a_1 - B\mu_2)P_2(a_1, B) \tag{3.14}$$

It should be noted that when $\Sigma_1 = \Sigma_2$ we always have the single root case, and one can verify that

$$B = (\mu_1 - \mu_2)^T (\Sigma_1 + \Sigma_2)^{-1} \tag{3.15}$$

satisfies $\frac{\partial g}{\partial B} = 0$. This suggests that one should start the
minimization procedure using this B as an initial guess, even
when $\Sigma_1 \neq \Sigma_2$.

4. NUMERICAL RESULTS

The procedure presented in Section 3 was initially im-
plemented on the Univac 1108 using a Davidon-Fletcher-Powell
minimization program (SUBROUTINE DAVIDN) provided by the NASA/
Johnson Space Center, Houston, Texas. A subsequent version of
the program was developed for the IBM 360 using a Fletcher-
Powell minimization program (SUBROUTINE DFMFP) from the IBM
Scientific Subroutine Package. In both programs, the starting
vector for the minimization procedure is computed using formula
(3.15).

The mean vectors and covariance matrices for Runs 1-3
(on the following pages) were computed from data appearing
in Fisher (1936). For Runs 4-5, we used the Standard Data 1-4
(herein labeled Fukunaga I-IV) from (Fukunaga (1972), pp. 46-47).
Three Sheets are given for each Run. The first (Sheet A) contains the class statistics, the minimizing B, associated transformed statistics, and the probability of misclassification for
B. Sheet B displays the graphs of the transformed density
functions and the associated decision regions. Sheet C presents
the results obtained from classifying observation vectors
randomly generated using the class statistics and the decision
regions determined by B.

The data generation and classification program was written
by Mrs. Louise Darcey. The graphs were made on a Hewlitt-Packard
9100B by Professor Jack Bryant. We are grateful to both for
their support.

CLASS 1 = Setosa

CLASS 2 = Versicolor

$$\Sigma_1 = \begin{pmatrix} .122 & .097 & .016 & .010 \\ .097 & .141 & .010 & .009 \\ .016 & .010 & .030 & .006 \\ .010 & .009 & .006 & .011 \end{pmatrix}$$

$$\Sigma_2 = \begin{pmatrix} .261 & .083 & .179 & .055 \\ .083 & .096 & .081 & .040 \\ .179 & .081 & .216 & .072 \\ .055 & .040 & .072 & .038 \end{pmatrix}$$

$$\mu_1 = \begin{pmatrix} 5.006 \\ 3.428 \\ 1.462 \\ 0.246 \end{pmatrix} \qquad \mu_2 = \begin{pmatrix} 5.936 \\ 2.770 \\ 4.260 \\ 1.326 \end{pmatrix}$$

$$B = (0.022, 0.383, -0.548, -0.743)$$

$$B\mu_1 = .440 \qquad\qquad B\mu_2 = -2.126$$

$$B\Sigma_1 B^T = 0.031 \qquad\qquad B\Sigma_2 B^T = 0.097$$

$$g(B) = 0.717 \times 10^{-7}$$

Run 1 - Sheet A

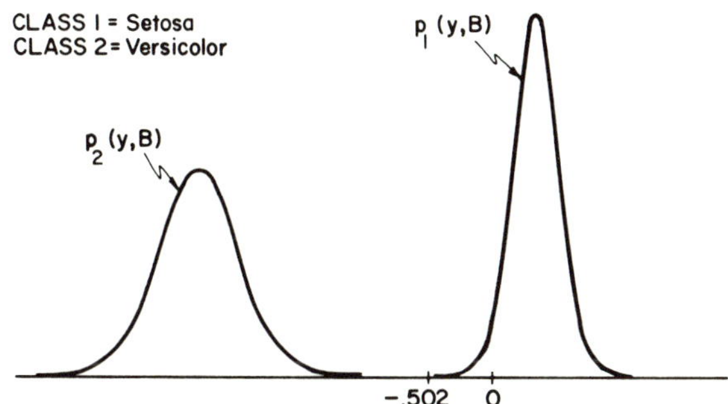

$$R_1(B) = [-0.502, 3.843]$$
$$R_2(B) = (-\infty, -0.502) \cup (3.843, +\infty)$$

Run 1 - Sheet B

CLASS 1 = Setosa

CLASS 2 = Versicolor

NUMBER OF RANDOM VECTORS GENERATED USING CLASS 1 STATISTICS = 200

NUMBER CORRECTLY CLASSIFIED = 200

NUMBER INCORRECTLY CLASSIFIED = 0

NUMBER OF RANDOM VECTORS GENERATED USING CLASS 2 STATISTICS = 200

NUMBER CORRECTLY CLASSIFIED = 200

NUMBER INCORRECTLY CLASSIFIED = 0

PERCENT INCORRECTLY CLASSIFIED = 0.00%

Run 1 - Sheet C

CLASS 1 = Setosa

CLASS 2 = Virginica

$$\Sigma_1 = \begin{pmatrix} .122 & .097 & .016 & .010 \\ .097 & .141 & .010 & .009 \\ .016 & .010 & .030 & .006 \\ .010 & .009 & .006 & .011 \end{pmatrix}$$

$$\Sigma_2 = \begin{pmatrix} .396 & .092 & .297 & .048 \\ .092 & .102 & .070 & .047 \\ .297 & .070 & .298 & .048 \\ .048 & .047 & .048 & .074 \end{pmatrix}$$

$$\mu_1 = \begin{pmatrix} 5.006 \\ 3.428 \\ 1.462 \\ 0.246 \end{pmatrix} \qquad \mu_2 = \begin{pmatrix} 6.588 \\ 2.974 \\ 5.552 \\ 2.026 \end{pmatrix}$$

$$B = (0.285, 0.216, -0.658, -0.662)$$

$$B\mu_1 = 1.046 \qquad\qquad B\mu_2 = -2.470$$

$$B\Sigma_1 B^T = 0.036 \qquad\qquad B\Sigma_2 B^T = 0.089$$

$$g(B) = 0.230 \times 10^{-12}$$

Run 2 - Sheet A

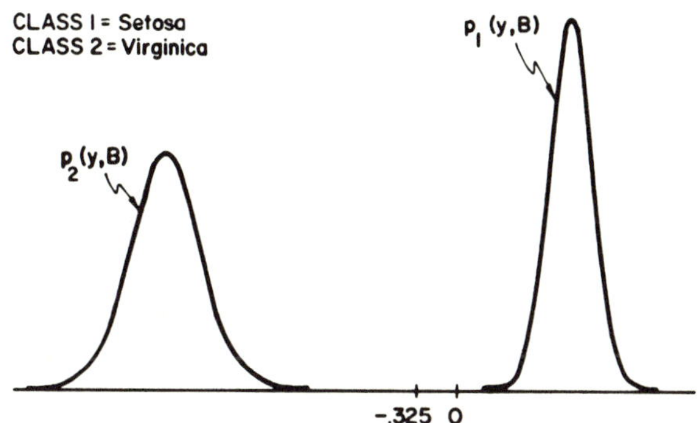

$$R_1(B) = [-0.325, 7.131]$$
$$R_2(B) = (-\infty, -0.325) \cup (7.131, +\infty)$$

Run 2 - Sheet B

CLASS 1 = Setosa

CLASS 2 = Virginica

NUMBER OF RANDOM VECTORS GENERATED USING CLASS 1 STATISTICS = 200

NUMBER CORRECTLY CLASSIFIED = 200

NUMBER INCORRECTLY CLASSIFIED = 0

NUMBER OF RANDOM VECTORS GENERATED USING CLASS 2 STATISTICS = 200

NUMBER CORRECTLY CLASSIFIED = 200

NUMBER INCORRECTLY CLASSIFIED = 0

PERCENT INCORRECTLY CLASSIFIED = 0.00%

Run 2 - Sheet C

CLASS 1 = Versicolor

CLASS 2 = Virginica

$$\Sigma_1 = \begin{pmatrix} .261 & .083 & .179 & .055 \\ .083 & .097 & .081 & .040 \\ .179 & .081 & .216 & .072 \\ .055 & .040 & .072 & .038 \end{pmatrix}$$

$$\Sigma_2 = \begin{pmatrix} .396 & .092 & .297 & .048 \\ .092 & .102 & .070 & .047 \\ .297 & .070 & .299 & .048 \\ .048 & .047 & .048 & .074 \end{pmatrix}$$

$$\mu_1 = \begin{pmatrix} 5.963 \\ 2.770 \\ 4.260 \\ 1.326 \end{pmatrix} \qquad \mu_2 = \begin{pmatrix} 6.588 \\ 2.974 \\ 5.552 \\ 2.026 \end{pmatrix}$$

$$B = (0.217, \ 0.360, \ -0.431, \ -0.799)$$

$$B\mu_1 = -0.607 \qquad\qquad B\mu_2 = -1.508$$

$$B\Sigma_1 B^T = 0.051 \qquad\qquad B\Sigma_2 B^T = 0.061$$

$$g(B) = 0.028$$

Run 3 - Sheet A

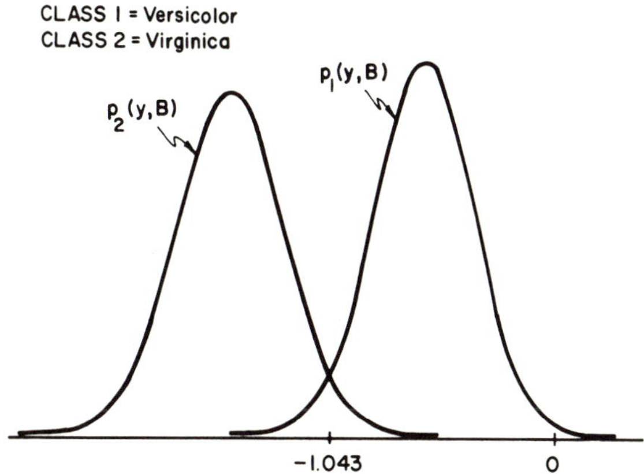

$$R_1(B) = [-1.043, 9.026]$$
$$R_2(B) = (-\infty, -1.043) \cup (9.026, +\infty)$$

Run 3 - Sheet B

CLASS 1 = Versicolor

CLASS 2 = Virginica

NUMBER OF RANDOM VECTORS GENERATED USING CLASS 1 STATISTICS = 200

NUMBER CORRECTLY CLASSIFIED = 193

NUMBER INCORRECTLY CLASSIFIED = 7

NUMBER OF RANDOM VECTORS GENERATED USING CLASS 2 STATISTICS = 200

NUMBER CORRECTLY CLASSIFIED = 194

NUMBER INCORRECTLY CLASSIFIED = 6

PERCENT INCORRECTLY CLASSIFIED = 3.25%

Run 3 - Sheet C

CLASS 1 = Fukunaga I

CLASS 2 = Fukunaga II

$$\Sigma_1 = \begin{pmatrix} 1.034 & 1.281 & 0.351 & -0.293 & 0.098 & 0.301 & 0.141 & 1.336 \\ 1.281 & 1.967 & 0.664 & -0.219 & 0.259 & 0.556 & 0.276 & 2.094 \\ 0.351 & 0.664 & 7.138 & 1.192 & 2.726 & 1.116 & 0.678 & 2.097 \\ -0.293 & -0.219 & 1.192 & 2.269 & 1.367 & 0.146 & 0.201 & -0.308 \\ 0.098 & 0.259 & 2.726 & 1.367 & 5.727 & 1.280 & 0.933 & 2.107 \\ 0.301 & 0.556 & 1.116 & 0.146 & 1.280 & 2.941 & 1.949 & 2.197 \\ 0.141 & 0.276 & 0.678 & 0.201 & 0.933 & 1.949 & 1.577 & 1.229 \\ 1.336 & 2.094 & 2.097 & -0.308 & 2.107 & 2.197 & 1.229 & 6.606 \end{pmatrix}$$

$$\Sigma_2 = \begin{pmatrix} 4.792 & 4.417 & 4.244 & 2.406 & 1.798 & 0.790 & 0.785 & 2.993 \\ 4.417 & 5.074 & 4.636 & 2.798 & 1.824 & 0.639 & 0.644 & 2.799 \\ 4.244 & 4.636 & 5.428 & 3.224 & 2.111 & 0.903 & 1.131 & 2.943 \\ 2.406 & 2.798 & 3.224 & 5.287 & 3.006 & 1.326 & 1.897 & 2.648 \\ 1.798 & 1.824 & 2.111 & 3.006 & 3.574 & 2.229 & 2.471 & 1.915 \\ 0.790 & 0.639 & 0.903 & 1.326 & 2.229 & 4.008 & 2.405 & 1.106 \\ 0.785 & 0.644 & 1.131 & 1.897 & 2.471 & 2.405 & 4.507 & 1.727 \\ 2.993 & 2.799 & 2.943 & 2.648 & 1.915 & 1.106 & 1.727 & 3.972 \end{pmatrix}$$

$$\mu_1 = \begin{pmatrix} 7.825 \\ 6.750 \\ 5.835 \\ 8.525 \\ 6.615 \\ 7.065 \\ 7.865 \\ 4.435 \end{pmatrix} \qquad \mu_2 = \begin{pmatrix} 5.760 \\ 5.715 \\ 5.705 \\ 4.150 \\ 6.225 \\ 6.960 \\ 6.750 \\ 3.910 \end{pmatrix}$$

$$B = (0.710, \ -0.492, \ -0.113, \ 0.457, \ -0.135, \ -0.056, \ 0.102, \ 0.027)$$

$$B\mu_1 = 5.099 \qquad\qquad B\mu_2 = 2.089$$

$$B\Sigma_1 B^T = 0.476 \qquad\qquad B\Sigma_2 B^T = 1.386$$

$$g(B) = 0.051$$

Run 4 - Sheet A

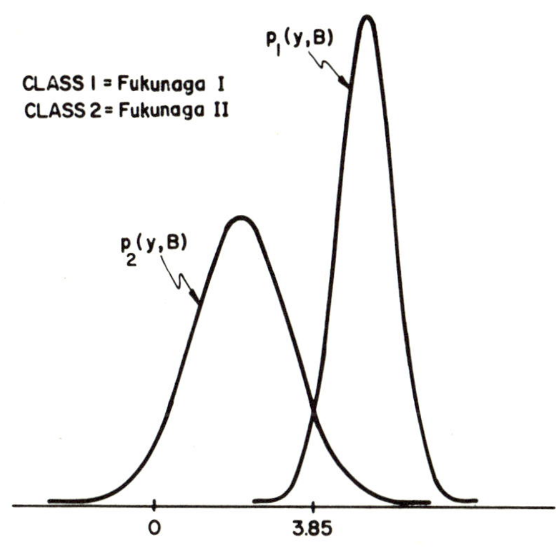

$$R_1(B) = [3.85, 9.50]$$
$$R_2(B) = (-\infty, 3.85) \cup (9.50, +\infty)$$

Run 4 - Sheet B

CLASS 1 = Fukunaga I

CLASS 2 = Fukunaga II

NUMBER OF RANDOM VECTORS GENERATED USING CLASS 1 STATISTICS = 200

NUMBER CORRECTLY CLASSIFIED = 194

NUMBER INCORRECTLY CLASSIFIED = 6

NUMBER OF RANDOM VECTORS GENERATED USING CLASS 2 STATISTICS = 200

NUMBER CORRECTLY CLASSIFIED = 184

NUMBER INCORRECTLY CLASSIFIED = 16

PERCENT INCORRECTLY CLASSIFIED = 5.50%

Run 4 - Sheet C

CLASS 1 = Fukunaga III
CLASS 2 = Fukunaga IV

$$\Sigma_1 = \begin{pmatrix} 1.638 & 2.153 & 1.482 & 1.695 & -0.557 & -2.443 & -0.710 & 1.983 \\ 2.153 & 3.596 & 2.461 & 2.436 & -0.591 & -3.711 & -0.493 & 2.434 \\ 1.482 & 2.461 & 2.500 & 2.834 & -0.665 & -2.621 & 0.248 & 1.738 \\ 1.695 & 2.436 & 2.834 & 4.704 & -0.629 & -2.913 & 0.576 & 2.471 \\ -0.557 & -0.591 & -0.665 & -0.629 & 19.000 & 0.896 & 8.622 & -0.254 \\ -2.443 & -3.711 & -2.621 & -2.913 & 0.896 & 5.856 & 1.357 & -2.915 \\ -0.710 & -0.493 & 0.248 & 0.576 & 8.622 & 1.357 & 20.800 & -0.622 \\ 1.983 & 2.434 & 1.738 & 2.471 & -0.254 & -2.915 & -0.622 & 3.214 \end{pmatrix}$$

$$\Sigma_2 = \begin{pmatrix} 5.116 & 4.726 & 4.058 & 1.821 & 1.109 & 1.289 & 1.029 & 2.232 \\ 4.736 & 5.684 & 4.523 & 2.311 & 1.273 & 1.328 & 1.151 & 2.425 \\ 4.058 & 4.523 & 6.117 & 2.525 & 1.321 & 1.501 & 1.274 & 2.191 \\ 1.821 & 2.311 & 2.525 & 4.432 & 2.481 & 2.179 & 1.080 & 1.784 \\ 1.109 & 1.273 & 1.321 & 2.481 & 2.134 & 2.325 & 1.017 & 1.030 \\ 1.289 & 1.328 & 1.501 & 2.179 & 2.325 & 4.099 & 2.019 & 1.803 \\ 1.029 & 1.151 & 1.274 & 1.080 & 1.017 & 2.019 & 1.872 & 2.081 \\ 2.232 & 2.425 & 2.191 & 1.784 & 1.030 & 1.803 & 2.081 & 3.806 \end{pmatrix}$$

$$\mu_1 = \begin{pmatrix} 6.610 \\ 5.060 \\ 5.980 \\ 3.975 \\ 9.020 \\ 14.685 \\ 10.640 \\ 4.175 \end{pmatrix} \qquad \mu_2 = \begin{pmatrix} 6.120 \\ 6.285 \\ 5.850 \\ 4.365 \\ 6.340 \\ 4.675 \\ 6.260 \\ 4.440 \end{pmatrix}$$

$$B = (0.406, -0.229, 0.061, -0.059, -0.057, 0.876, 0.066, -0.021$$

$$B\mu_1 = 14.620 \qquad\qquad B\mu_2 = 5.199$$

$$B\Sigma_1 B^T = 4.557 \qquad\qquad B\Sigma_2 B^T = 3.734$$

$$g(B) = .010$$

Run 5 - Sheet A

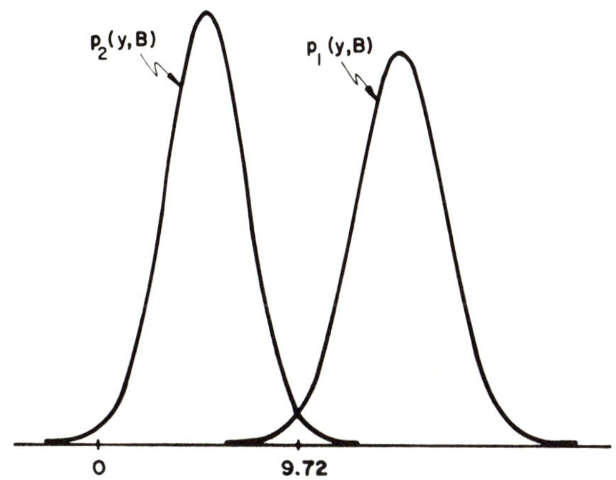

Run 5 - Sheet B

CLASS 1 = Fukunaga III

CLASS 2 = Fukunaga IV

NUMBER OF RANDOM VECTORS GENERATED USING CLASS 1 STATISTICS = 200

NUMBER CORRECTLY CLASSIFIED = 199

NUMBER INCORRECTLY CLASSIFIED = 1

NUMBER OF RANDOM VECTORS GENERATED USING CLASS 2 STATISTICS = 200

NUMBER CORRECTLY CLASSIFIED = 198

NUMBER INCORRECTLY CLASSIFIED = 2

PERCENT INCORRECTLY CLASSIFIED = 0.75%

Run 5 - Sheet C

5.   CONCLUDING REMARKS

The use of unequal (known) *a priori* population probabilities and cost functions is being incorporated into the existing
computational procedure. For the case $k = 1$, investigations
are underway to develop the theory and associated computational
procedure for the multi-population problem. It would be of
interest to obtain a starting vector, analogous to that of
(3.15), for the multi-population case.

Finally, we note the possibility of developing a "suboptimal
method", using the computational procedure discussed herein,
which determines a $k \times n$ matrix B, one row at a time, so as
to "nearly" minimize the probability of misclassification in
k-dimensional space $(1 < k < n)$. The theory for such a procedure has not been developed, even in the case of two
populations.

ACKNOWLEDGMENTS

L. F. Guseman, Jr. was a NASA-ASEE Summer Faculty Fellow
with the Earth Observations Division, NASA/Johnson Space Center,
Houston, Texas, during the preparation of this work. Homer F.
Walker was supported under NASA Contract NAS-9-12777 with the
University of Houston during the preparation of this work.

BIBLIOGRAPHY

Anderson, T. W.   (1958). *An Introduction to Multivariate
Statistical Analysis*. New York: John Wiley & Sons,
Inc.

Anderson, T. W. and Bahadur, R. R.   (1962). Classification
into two multivariate normal distributions with different covariance matrices. *Ann. Math. Statist.* 33,
420-31.

Fisher, R. A.   (1936). The use of multiple measurements
in taxonomic problems. *Ann. Eugenics* 7, 179-88.

Fukunaga, Keinosuke (1972). *Introduction to Statistical Pattern Recognition*. New York: Academic Press.

Guseman, L. F., Jr., and Walker, Homer F. (1973). On minimizing the probability of misclassification for linear feature selection. Submitted for publication.

Guseman, L. F., Jr. and Walker, Homer F. (1973). The differentiability of the probability of misclassification as a function of a linear feature selection matrix. Report #28, NASA Contract NAS-9-12777, Dept. of Math., Univ. of Houston.

Luenberger, D. G. (1969). *Optimization by Vector Space Methods*. New York: John Wiley and Sons, Inc.

A CLASS OF NONPARAMETRIC CLASSIFICATION RULES

Kamal C. Chanda

Jack Chao-sheng Lee

Wright State University
Dayton, Ohio

1. INTRODUCTION

Let F_1, F_2 be two cumulative distribution functions (cdf)
which are partly or completely unknown except for random samples
$\underset{\sim}{X} = (X_1, \ldots, X_{n_1})$, $\underset{\sim}{Y} = (Y_1, \ldots, Y_{n_2})$ drawn from F_1 and
F_2, respectively. A third random sample $\underset{\sim}{Z} = (Z_1, \ldots, Z_k)$ of
size k is available and the problem is to identify the cdf
F to which $\underset{\sim}{Z}$ belongs with one (or none) of the cdf's F_1,
F_2. This is the simplest type of classification problem and
has been discussed in great detail by authors such as Anderson
(1958) and Rao (1965), among others. If the forms of F_1, F_2
are known but otherwise F_1, F_2 involve some finite number of
unknown parameters, the usual rule is to replace the unknown
parameters by suitable well-behaved consistent estimators. The
sampling properties of these classification rules have been
partially studied by Anderson (1958), Cochran and Hopkins (1961),
Lachenbruch, Sneeringer and Revo (1973) and Sorum (1973).

These classification rules, with or without appropriate
parametric estimation, have near optimal properties (in an
appropriate sense) when the cdf's are normal. Properties of
similar classification rules when the cdf's are not normal
are investigated in Lachenbruch, et al. (1973). However, in
view of the fact that nonparametric test procedures have been
in use for quite some time and have been found to be extremely

effective, it seems logical to use appropriately defined non-parametric classification rules which hopefully should prove equally effective. It appears that the first significant attempt for such a classification rule is made in Hudimoto (1964), although, in a somewhat incidental way it also appears in the works of Das Gupta (1964), Fix and Hodges (1952),(1959) and Stoller (1954). The results derived in these articles, however, could not have gone far because of lack of appropriate mathematical tools which have been developed only in the last few years. Govindarajulu and Gupta (1972) use some of these mathematical tools to derive some interesting results concerning a class of nonparametric rules. Their results are, however, substantially different from what is being considered in this article.

The main purpose of this article is to make an effort in the direction of defining a series of nonparametric classification rules which would prove to be effective both in small samples when n_1, n_2 are small, and in large samples when n_1, n_2 both are large (we assume k to be usually finite). We consider one such nonparametric rule (a slight modification of Hudimoto's rule) in this article and investigate its properties in some detail. Subsequent articles will deal with other nonparametric classification rules.

Once we define a nonparametric rule, one way of assessing the efficiency of the rule is to compare it against the optimal (in a certain sense) rule when F_1, F_2 both are completely known or partially known except for some parameters which can be efficiently estimated.

2. SOME EFFICIENT PARAMETRIC CLASSIFICATION RULES

Given $\underset{\sim}{Z}$, $\underset{\sim}{X}$, $\underset{\sim}{Y}$, with known F_1, F_2 we define as follows a rule M (see Rao (1965)) of choosing one of the decisions $d_1: F = F_1$ and $d_2: F = F_2$. We assume that F_1, F_2, F are

all absolutely continuous and the probability density functions
(pdf) of F_1, F_2 are denoted by f_1, f_2, respectively. Then
we have Rule M:

$$\text{choose } d_1 \text{ if } \underset{\sim}{z} \in A,$$

$$\text{choose } d_2 \text{ if } \underset{\sim}{z} \in A^C \qquad (2.1)$$

where

$$A = \left\{ \underset{\sim}{z} : \prod_1^k f_1(z_j) \geq \prod_1^k f_2(z_j) \right\} \qquad (2.2)$$

If f_1, f_2 are completely unknown we may replace them by
some appropriate estimates $\hat{f}_1$ and $\hat{f}_2$, respectively, and
use the alternate Rule $\hat{M}$:

$$\text{choose } d_1 \text{ if } \underset{\sim}{z} \in \hat{A},$$

$$\text{choose } d_2 \text{ if } \underset{\sim}{z} \in \hat{A}^C \qquad (2.3)$$

where

$$\hat{A} = \left\{ \underset{\sim}{z} : \prod_1^k \hat{f}_1(z_j) \geq \prod_1^k \hat{f}_2(z_j) \right\} \qquad (2.4)$$

In particular, if f_i $(i = 1, 2)$ are known except for some
unknown parametric vector $\underset{\sim}{\theta}_i$, then we write $\hat{f}_i(z) = f(z, \hat{\underset{\sim}{\theta}}_i)$
where $\hat{\underset{\sim}{\theta}}_i$ are efficient estimators of $\underset{\sim}{\theta}_i$.

Estimation of f_i, parametrically or nonparametrically,
is an involved process and using it to define $\hat{M}$ may lead to
considerable loss in efficiency relative to M. A logical way
of defining $\hat{M}$ is, of course, to make sure that

$$P_1 = E_1[I_{\hat{A}}(\underset{\sim}{Z})] \qquad (2.5)$$

and

$$P_2 = E_2[I_{\hat{A}^C}(\underset{\sim}{Z})] \qquad (2.6)$$

where E_i is expectation with respect to F_i $(i = 1, 2)$ and
$I_{\hat{A}}$ is the indicator function of A, are both as large as

possible. This approach is taken by Sorum (1973). We shall not go into the merits of the rule M, especially when F_i involve the parameter vectors $\underset{\sim}{\theta}_i$ which are efficiently estimated, but simply assume that $\hat{M}$ is a reasonably efficient classification rule, at least for large n_1, n_2.

An entirely different type of nonparametric classification rule is now introduced.

3. A SIMPLE CLASSIFICATION RULE BASED ON RANKS

If the cdf's F_1, F_2 are such that on an average the Y-observations are likely to be larger than the X-observations and if the cdf's do not have very heavy tails, then one probably can assume that the X- and Y-observations are reasonably separated. If $F = F_1$, then the X- and Z-observations are fairly well mixed, and therefore, the latter are also separated from the Y-observations. Hence, on an average the total number of X- and Y-observations which are less than the Z-observations is smaller probably than if $F = F_2$. This is the intuitive basis of the rule M* described below.

Let

$$\phi(u) = \begin{cases} 0 & \text{if } u > 0 \\ \frac{1}{2} & \text{if } u \leq 0 \end{cases} \tag{3.1}$$

$$\psi(u) = 1 - 2\phi(u)$$

$$= \begin{cases} 1 & \text{if } u > 0 \\ 0 & \text{if } u \leq 0 \end{cases} \tag{3.2}$$

Define

$$W = \sum_{i=1}^{n_1} \sum_{j=1}^{n_2} \psi(X_i - Y_j)/n_1 n_2 \tag{3.3}$$

$$U_1 = \sum_{i=1}^{n_1} \sum_{j=1}^{k} \phi(X_i - Z_j) \tag{3.4}$$

$$U_2 = \sum_{i=1}^{n_2} \sum_{j=1}^{k} \phi(Y_i - Z_j) \tag{3.5}$$

$$V = U_1/kn_1 + U_2/kn_2 \tag{3.6}$$

Note that

$$n_1 n_2 W = \text{number of } (X_i, Y_j) \, (1 \le i \le n_1, 1 \le j \le n_2) \text{ with } X_i > Y_j \tag{3.7}$$

$$2U_1 = \text{number of } (X_i, Z_j) \, (1 \le i \le n_1, 1 \le j \le k) \text{ with } X_i \le Z_j \tag{3.8}$$

$$2U_2 = \text{number of } (Y_i, Z_j) \, (1 \le i \le n_2, 1 \le j \le k) \text{ with } Y_i \le Z_j \tag{3.9}$$

If $F = F_1 > F_2$, then

$$E[F_2(Z_1)] < E[F_1(Z_1)] = \frac{1}{2} \tag{3.10}$$

which implies that when $F = F_1 > F_2$

$$E(U_1/kn_1) = E[\phi(X_1 - Z_1)] = \frac{1}{2} E[F_1(Z_1)] = \frac{1}{4} \tag{3.11}$$

$$E(U_2/kn_2) = E[\phi(Y_1 - Z_1)] = \frac{1}{2} E[F_2(Z_1)] < \frac{1}{4} \tag{3.12}$$

$$E(V) < \frac{1}{2} \tag{3.13}$$

$$E(W) < \frac{1}{2} \tag{3.14}$$

Similarly if

$$F_1 > F_2 = F \tag{3.15}$$

$$E(V) > \frac{1}{2} \tag{3.16}$$

If $F_2 > F_1$ the inequalities in (3.14) will be reversed.
Therefore a reasonable rule M^* can be defined as

$$\text{choose } d_1 \text{ if } (\underset{\sim}{X}, \underset{\sim}{Y}, \underset{\sim}{Z}) \in A^*$$

$$\text{choose } d_2 \text{ if } (\underset{\sim}{X}, \underset{\sim}{Y}, \underset{\sim}{Z}) \in A^{*C} \tag{3.17}$$

where

$$A^* = \{x, y, z\}: \ w > \frac{1}{2}, \ v > \frac{1}{2} \ \text{ or } \ w \leq \frac{1}{2}, \ v \leq \frac{1}{2}\} \qquad (3.18)$$

Here x, y, z, v, w are the respective values of X, Y, Z, V and W.

4. SAMPLING PROPERTIES OF $\hat{M}$ AND M^*

To the best of these authors' knowledge no systematic efforts have yet been made to study the sampling properties of nonparametric classification rules. Mention may be made in this connection of the work by Hudimoto (1964). Here he investigates some approximate bounds for the probabilities of correct classification for a particular class of nonparametric rules similar to (but much more restrictive than) those discussed here.

The present investigation consists of two parts: (1) derivation of the asymptotic (i.e., when $n_1, n_2 \to \infty$) properties of M^* together with those of $\hat{M}$, with special reference to the situation where $F_j(x) = G(x - \theta_j)$, $j = 1, 2,$ and $F(x) = G(x - \theta)$, for some cdf G and θ, θ_1, θ_2, are arbitrary values of a scalar parameter (usually we assume that $\theta = \theta_1$ or θ_2, although the results are given in terms of arbitrary θ); and (2) simulation study of the small sample properties of M^* and $\hat{M}$ for the particular case when $F_j(x) = G(x - \theta_j)$, $F = F_1$ or F_2, and G is either normal, logistic or Cauchy. As can be seen quite easily the various classification probabilities relating to M^* are identifiable with the powers of a Wilcoxon-like test against specified alternatives and as such are somewhat tedious to compute. Similar remarks hold for $\hat{M}$. Therefore, the authors feel that such computation should be done through reasonably efficient simulation procedures which, hopefully, will provide enough information about the sampling properties of M^* and M when n_1, n_2, k are small. Accordingly we have done so.

4.1. Asymptotic Properties Of M*

The results in this section are derived without any
assumption about F, F_1, F_2 $(F_1 \neq F_2)$. Thus, in particular
F may be equal to F_1 or F_2 but in general F may not be
so.

Consider the rule M^* and let

$$Q_k = P(\text{choose } \delta_1)$$

$$= P(A^*) \tag{4.1}$$

where A^* is defined in (3.17). Since F_1, F_2 are assumed
fixed (although arbitrary) we shall regard Q_k as a function
of F only. If $Q_k(j)$ denotes the probability of correct
classification to F_j, then

$$Q_k(1) = Q_k \quad \text{when} \quad F = F_1 \tag{4.2}$$

$$Q_k(2) = 1 - Q_k \quad \text{when } F = F_2 \tag{4.3}$$

We now state the following theorem of which the proof is
given in the Appendix.

__Theorem 4.1.__ Let F, F_1, F_2 be absolutely continuous cdf's
and let Q_k be as defined in (4.1). Let

$$\pi = \frac{1}{2} - \nu \tag{4.4}$$

$$\nu = E(F_2(X_1)) \tag{4.5}$$

Assume that $\left|F'(z)\right|$, $\left|F_1'(z)\right|$, $\left|F_2'(z)\right| < N$ for all z and
for some $N > 0$. From here on we use N as a generic symbol
to denote a finite positive quantity which is independent of
n_1 and n_2.

Then as n_1, $n_2 \to \infty$

$$Q_k = \begin{cases} P\left[\sum_{j=1}^{k} \chi(Z_j) < 0\right] + o(n^{-\alpha}) & \text{if } \pi \geq 0 \\[4mm] P\left[\sum_{j=1}^{k} \chi(Z_j) > 0\right] + o(n^{-\alpha}) & \text{if } \pi < 0 \end{cases} \tag{4.6}$$

where $\alpha < \frac{1}{2}$, $n = n_1 n_2/(n_1+n_2)$, $Z_1, \ldots, Z_k$ are as defined earlier and

$$\chi(z) = F_1(z) + F_2(z) - 1 \tag{4.7}$$

If we denote the right-hand side probability in (4.2) by Q_k^0 then

$$\lim_{n \to \infty} Q_k = Q_k^0 \tag{4.8}$$

4.1.1. A Special Case $F(x) = G(x-\theta)$, $F_j(x) = G(x-\theta_j)$, $j = 1, 2$

We now consider the special case when $F(x) = G(x-\theta)$, $F_j(x) = G(x-\theta_j)$, $j = 1, 2$. Let $\Delta = (\theta_2-\theta_1)/2$. Then it is easy to see that

$$Q_k = \begin{cases} P\left[\sum\limits_{j=1}^{k} T(W_j) < 0\right] & \text{if } \pi \geq 0 \\[2ex] P\left[\sum T(W_j) \geq 0\right] & \text{if } \pi > 0 \end{cases} \tag{4.9}$$

where

$$T(w) = G(w+\Delta) + G(w-\Delta) - 1 \tag{4.10}$$

and $W_1, \ldots, W_k$ are independent and identically distributed (iid) random variables (r.v.) with a common cdf H where $H(x) = G(x+\delta)$, $\delta = (\theta_1+\theta_2)/2 - \theta$. Note here that $\pi \geq 0 \iff \Delta \geq 0$. Also when $F = F_1$, $\delta = \Delta$ and when $F = F_2$, $\delta = -\Delta$. If $G(x)$ is symmetric about $x = 0$, then for all w

$$T(w) = G(w+\Delta) - G(-w+\Delta) \tag{4.11}$$

$$T'(w) = g(w+\Delta) + g(-w+\Delta) > 0 \tag{4.12}$$

$$g(w) = G'(w) \tag{4.13}$$

from which it follows that $T(w) \uparrow$ in w and for all $w \geq 0$

$$T(-w) = -T(w) \tag{4.14}$$

Consequently, when $\Delta \geq 0$

$$Q_1^0 = P[T(W_1)<0] = P(W_1<0) = G(\delta) \tag{4.15}$$

$$Q_2^0 = P[T(W_1)+T(W_2)<0]$$

$$= P[T(W_1)<T(-W_2)]$$

$$= P(W_1<-W_2)$$

$$= P(W_1+W_2<0) = G_2(2\delta) \tag{4.16}$$

Similarly, for $\Delta < 0$ $Q_1^0 = 1-G(\delta)$ and $Q_2^0 = 1-G_2(2\delta)$ where $G_2 = G*G$. We are unable to prove or disprove whether for any k, $Q_k^0 = G_k(k\delta)$, $G_k = G*G*\ldots*G$ = the kth convolution of G.

For the optimal rule M defined in (2.1) we have for the special case considered above

$$P_k^0 = P(\text{choose } d_1)$$

$$= P\left[\sum_{j=1}^{k} h(W_j)<0\right] \tag{4.17}$$

where

$$h(w) = \log g(w-\Delta) - \log g(w+\Delta) \tag{4.18}$$

Suppose now we assume that $G(w)$ is symmetric about $w = 0$ and $g(w) = G'(w)$ is monotone nondecreasing for $w < 0$ and monotone nonincreasing for $w > 0$. First let $\Delta > 0$. Then $w \geq 0$ implies that $|w-\Delta| \leq |w| + |\Delta| = w + \Delta$, $\log g(w-\Delta)$ $= \log g(|w-\Delta|) \geq \log g(w+\Delta)$, $h(w) \geq 0$. Again $w < 0$ implies that $|w+\Delta| \leq |w| + |\Delta| = -w + \Delta$, $\log g(w+\Delta) = \log g(|w+\Delta|) \geq \log g(-w+\Delta)$ $= \log g(w-\Delta)$, i.e., $h(w) < 0$. Therefore $w < 0$ is equivalent to $h(w) < 0$ and $w \geq 0$ is equivalent to $h(w) \geq 0$. This holds even when $\Delta < 0$ as we can see by simply replacing w by $-w$ and using the same argument as above. Hence

$$P_1^0 = Q_1^0 . \tag{4.19}$$

The result (4.6) holds true for normal, logistic and Cauchy.

4.2. Asymptotic Properties of $\hat{M}$

Consider the rule $\hat{M}$ and let

$$P_k = P(\text{choose } d_1)$$

$$= P(\hat{A}) \tag{4.20}$$

where $\hat{A}$ is defined in (2.3). Since F_1, F_2 are assumed fixed (although arbitrary) we shall regard P_k as a function of F only. If $P_k(j)$ denotes the probability of correct classification to F_j $(j = 1, 2)$, then

$$P_k(1) = P_k \quad \text{when} \quad F = F_1 \tag{4.21}$$

$$P_k(2) = 1 - P_k \quad \text{when} \quad F = F_2 \tag{4.22}$$

We can state and prove a theorem similar to Theorem 4.1 concerning the asymptotic behavior of $\hat{M}$, but the exact conditions for such a theorem are somewhat messy. We simply state (without proof) that under certain moment conditions on $\log f_1(Z_1) - \log f_2(Z_1)$, and $\log \hat{f}_j(Z_1) - \log f_j(Z_1)$, $j = 1, 2$ it can be shown that

$$\left| P_k - P_k^0 \right| < Nm^{-r/(2r+1)} \tag{4.23}$$

for some $r > 0$, where P_k^0 is defined in (2.1) and $m = \min(n_1, n_2)$.

4.3. P_k vs Q_k and P_k^0 vs Q_k^0 for the Special Case of Section 4.1.1.

From the results (4.9) and (4.23) it follows that the rate of convergence of Q_k to Q_k^0 is about the same as the rate of convergence of P_k to P_k^0. However, it seems reasonable to believe (a formal proof is difficult to construct and we have not tried it here) that

$$P_k^0 \geq Q_k^0 \tag{4.24}$$

although (as we shall see below) that in some special cases

$$P_k^0 = Q_k^0 \tag{4.25}$$

so that in those cases it may be better to use the nonparametric rule M^* (because of its simplicity) in preference to $\hat{M}$. This, of course, is based on the presumption that M is better than $\hat{M}$ and M^*, which in small samples may not be true (at least we have not proved it so).

Note that for the special case considered here, namely, $F(x) = G(x-\theta)$, $F_j(x) = F(x-\theta_j)$, $j = 1$, 2, (with further assumptions that $G(x)$ is symmetric about $x = 0$, unimodal with $G'(x) = g(x)$ monotone nondecreasing for $x < 0$ and monotone nonincreasing for $x > 0$)

$$P_k^0 = P\left[\sum_1^k h(W_j) < 0\right] \tag{4.26}$$

$$Q_k^0 = P\left[\sum_1^k T(W_j) < 0\right] \tag{4.27}$$

$$h(w) = \log g(w-\Delta) - \log g(w+\Delta) \tag{4.28}$$

$$T(w) = G(w+\Delta) - G(-w+\Delta) \tag{4.29}$$

$$\Delta = (\theta_2 - \theta_1)/2 \tag{4.30}$$

and $Z_1, \ldots, Z_k$ are iid r.v. with cdf F. As we have seen in Section 4.1.1 $P_1^0 = Q_1^0$. For the special case when G is normal $N(0,1)$,

$$P_k^0 = \begin{cases} \Phi(\delta\sqrt{k}) & \text{if } \Delta > 0 \\ 1 - \Phi(\delta\sqrt{k}) & \text{if } \Delta < 0 \end{cases} \tag{4.31}$$

where $\delta = (\theta_1 + \theta_2)/2 - \theta$, $\Phi(x) = \int_{-\infty}^{x} \phi(y)\,dy$, $\phi(y) = e^{-\frac{1}{2}y^2}/\sqrt{2\pi}$.

On the other hand, we have the following theorem, of which the proof is given in the Appendix.

<u>Theorem 4.2.</u> Let G be normal $N(0,1)$. Let θ_1, θ_2 be estimated by $\overline{X}$ and $\overline{Y}$ respectively where

$$\overline{X} = n_1^{-1} \sum_{j=1}^{n_2} X_j, \quad \overline{Y} = n_2^{-1} \sum_{j=1}^{n_2} Y_j. \tag{4.32}$$

Define P_k as in (4.20). Then

$$\lim_{n\to\infty} n(P_k - P_k^0) = \begin{cases} -\delta k^{\frac{3}{2}} \; \phi(\delta\sqrt{k}) & \text{if } \Delta > 0 \\[2ex] \delta k^{\frac{3}{2}} \; \phi(\delta\sqrt{k}) & \text{if } \Delta < 0 \end{cases} \tag{4.33}$$

where $n = n_1 n_2/(n_1 + n_2)$ and δ, ϕ are defined above. If $\overline{X}$, $\overline{Y}$ are replaced by any consistent estimators of θ_1 and θ_2, respectively, then

$$P_k - P_k^0 = o(n^{-\frac{1}{2}}) . \tag{4.34}$$

Note that if G is normal $N(0,1)$, then from (4.12)

$$Q_2^0 = \begin{cases} G_2(2\delta) = \Phi(\delta\sqrt{2}) & \text{if } \Delta > 0 \\[1ex] 1 - G_2(2\delta) = 1 - \Phi(\delta\sqrt{2}) & \text{if } \Delta < 0 \end{cases}$$

$$= P_2^0 \tag{4.35}$$

Thus for the special case of normal distribution

$$Q_k^0 = P_k^0, \quad k = 1,\ 2 \tag{4.36}$$

and for other unimodal symmetric distribution, like logistic and Cauchy,

$$Q_1^0 = P_1^0 \tag{4.37}$$

5. SOME SIMULATION STUDIES CONCERNING SMALL SAMPLE BEHAVIOR OF M* AND $\hat{M}$

Discussions in Section 4 are entirely concerned with the asymptotic behavior of M* and $\hat{M}$. As briefly stated at the

end of Section 4, it is somewhat tedious to work out the exact
sampling properties of M^* and $\hat{M}$ when n_1, n_2 are small.
Therefore, we make an attempt in this section to resort to
some simulation methods which we hope will prove sufficiently
informative about the properties of M^* and $\hat{M}$ when n_1, n_2
are small. We exclusively confine ourselves to the computation
of $Q_k(1)$, $1-Q_k(2)$ and $P_k(1)$, $1-P_k(2)$. No attempt is made
to find out what happens when F is neither F_1 nor F_2, although only one out of the two decisions d_1 and d_2 is
chosen.

Three situations are investigated, namely, when (i) G
is normal $N(0,1)$, (ii) G is logistic with the location
parameter 0 and scale parameter chosen such that the variance
of the distribution is unity and (iii) G is standard Cauchy.
As usual $F_j(x) = G(x-\theta_j)$, $j = 1, 2$, and $F = F_1$ or F_2.
For each set of parameters $\theta_1, \theta_2, n_1, n_2, k_1$ (k_1 = size of the
sample $\underset{\sim}{Z}$ from F_1) we repeat the experiment independently
500 times and compute the proportions $\hat{P}_{k_1}(1)$ of correct
classification of $\underset{\sim}{Z}$ to F_1. Similarly, 500 repeated and
independent experiments are made for the same set of
parameters to compute the proportion $\hat{Q}_{k_1}(1)$. The experiment
is repeated exactly as above for the same set of parameters
$\theta_1, \theta_2, n_1, n_2, k_2$ (k_2 = the size of the sample Z from F_2)
to compute the proportions $\hat{P}_{k_2}(2)$ and $\hat{Q}_{k_2}(2)$. It is believed that since 500 is a large number, the proportions
adequately estimate the true probabilities $P_k(j)$ and
$Q_k(j)$, $j = 1, 2$. We provide the approximate 95% confidence
intervals for these probabilities, and the midpoints of these
intervals give us the estimated probabilities. The results
are displayed in Tables 1, 2, and 3 which follow. (In the
Tables PK1 is the same as $P_k(1)$ in the text, and so on).

Some further comments are necessary in order to explain
how we define the classification rule $\hat{M}$ in each individual
case.

For the normal distribution we choose the estimators $\bar{X}$, $\bar{Y}$ for θ_1 and θ_2, respectively. For the logistic situation we estimate θ_1 by

$$\hat{\theta}_1 = 6/n_1 (n_1+1)^2 \sum_{r=1}^{n_1} r(n_1+1-r) X_{r:n_1} \tag{5.1}$$

where $X_{1:n} \leq \ldots \leq X_{n:n}$ are the ordered observations $X_1, \ldots, X_n$. Similarly $\hat{\theta}_2$ can be defined. This estimate is believed to have optimal properties (see Gupta and Waknis (1964)). Finally, for the Cauchy distribution θ_1 is estimated by

$$\hat{\theta}_1 = c_1^{-1} \sum_{r=3}^{n_1-2} \cos\left(\frac{2r\pi}{n_1+1}\right) \left[\cos\left(\frac{2r\pi}{n_1+1}\right)-1\right] X_{r:n_1} \tag{5.2}$$

where

$$c_1 = n_1/2 - 2 \sum_{r=1}^{2} \cos\left(\frac{2r\pi}{n_1+1}\right)\left[\cos\left(\frac{2r\pi}{n_1+1}\right)-1\right] \tag{5.3}$$

Similarly we define $\hat{\theta}_2$. This estimator again is believed to have optimal properties (see Barnett (1966), Chernoff, Gastwirth and Johns (1967), and Jung (1955)).

On closely analyzing the data in the tables we can conclude generally as follows. (1) The probabilities of correct classification for fixed $|\theta_1-\theta_2|$ are of the same order for the normal and the logistic, but of significantly lower order for the Cauchy. This is true for both the rules $\hat{M}$ and M^*. (2) The probabilities of correct classification increases both with $|\theta_1-\theta_2|$ and k_1 or k_2. These conclusions are quite consistent with the properties of the three distributions. (3) The rules $\hat{M}$ and M^* provide approximately the same order of protection against misclassification. Therefore, it appears that for at least the situations where the distributions are unimodal and symmetric it is quite safe to use the nonparametric classification rule M^*.

TABLE 1   95% Confidence Intervals for the Probabilities of Correct Classification

(Normal Distribution with Mean θ and Variance 1)

A. $\theta_1 = -1$ $\theta_2 = 1$

N1	N2	K1	K2	PK1	QK1	PK2	QK2
10	10	1	1	(0.793,0.859)	(0.821,0.883)	(0.810,0.874)	(0.773,0.843)
10	10	1	3	(0.804,0.868)	(0.825,0.887)	(0.933,0.971)	(0.917,0.959)
10	10	1	5	(0,806,0.870)	(0.814,0.878)	(0.945,0.979)	(0.940,0.976)
10	10	3	1	(0.945,0.979)	(0.958,0.986)	(0.814,0.878)	(0.780,0.848)
10	10	3	3	(0.938,0.974)	(0.931,0.969)	(0.938,0.974)	(0.910,0.954)
10	10	3	5	(0.933,0.971)	(0.938,0.974)	(0.955,0.985)	(0.940,0.976)
10	10	5	1	(0.955,0.985)	(0.958,0.986)	(0.847,0.905)	(0.808,0.872)
10	10	5	3	(0.973,0.995)	(0.965,0.991)	(0.936,0.972)	(0.912,0.956)
10	10	5	5	(0.953,0.983)	(0.950,0.982)	(0.955,0.985)	(0.948,0.980)
10	15	1	1	(0.791,0.857)	(0.778,0.846)	(0.817,0.879)	(0.793,0.859)
10	15	1	3	(0.830,0.890)	(0.819,0.881)	(0.922,0.962)	(0.922,0.962)
10	15	1	5	(0.771,0.841)	(0.793,0.859)	(0.976,0.996)	(0.965,0.991)
10	15	3	1	(0.938,0.974)	(0.931,0.969)	(0.817,0.879)	(0.812,0.876)
10	15	3	3	(0.933,0.971)	(0.917,0.959)	(0.936,0.972)	(0.912,0.956)
10	15	3	5	(0.912,0.956)	(0.896,0.944)	(0.960,0.988)	(0.948,0.980)
10	15	5	1	(0.973,0.995)	(0.968,0.992)	(0.804,0.868)	(0.788,0.856)
10	15	5	3	(0.960,0.988)	(0.960,0.988)	(0.936,0.972)	(0.924,0.964)
10	15	5	5	(0.953,0.983)	(0.955,0.985)	(0.973,0.995)	(0.965,0.991)
15	10	1	1	(0.799,0.865)	(0.806,0.870)	(0.810,0.874)	(0.795,0.861)
15	10	1	3	(0.806,0.870)	(0.808,0.872)	(0.938,0.974)	(0.926,0.966)
15	10	1	5	(0.806,0.870)	(0.814,0.878)	(0.978,0.998)	(0,970,0.994)
15	10	3	1	(0.922,0.962)	(0.926,0.966)	(0.810,0.874)	(0.793,0.859)
15	10	3	3	(0.929,0.967)	(0.931,0.969)	(0.922,0.962)	(0.912,0.956)
15	10	3	5	(0.915,0.957)	(0.919,0.961)	(0.973,0.995)	(0.970,0.994)
15	10	5	1	(0.955,0.985)	(0.968,0.992)	(0.799,0.865)	(0.780,0.848)
15	10	5	3	(0.965,0.991)	(0.963,0.989)	(0.940,0.976)	(0.929,0.967)
15	10	5	5	(0.973,0.995)	(0.965,0.991)	(0.970,0.994)	(0.965,0.991)
15	15	1	1	(0,830,0.890)	(0.849,0.907)	(0.823,0.885)	(0.784,0.852)
15	15	1	3	(0.793,0.859)	(0,817,0.879)	(0.943,0.977)	(0.919,0.961)
15	15	1	5	(0.827,0.889)	(0.847,0.905)	(0.970,0.994)	(0.960,0.988)
15	15	3	1	(0.948,0.980)	(0.950,0.982)	(0.830,0.890)	(0.810,0.874)
15	15	3	3	(0.940,0.976)	(0.929,0.967)	(0.908,0.952)	(0.894,0.942)
15	15	3	5	(0.929,0.967)	(0.924,0.964)	(0.968,0.992)	(0.963,0.989)
15	15	5	1	(0.976,0.996)	(0.973,0.995)	(0.817,0.879)	(0.786,0.854)
15	15	5	3	(0.984,1.000)	(0.970,0.994)	(0.912,0.956)	(0.898,0.946)
15	15	5	5	(0.984,1.000)	(0.978,0.998)	(0.976,0.996)	(0.970,0.994)

TABLE 1 (continued)

B. $\theta_1 = -1$ $\theta_2 = 0.5$

N1	N2	K1	K2	PK1	QK1	PK2	QK2
10	10	1	1	(0.731,0.805)	(0.763,0.833)	(0.714,0.790)	(0.685,0.763)
10	10	1	3	(0.735,0.809)	(0.742,0.814)	(0.863,0.917)	(0.845,0.903)
10	10	1	5	(0.723,0.797)	(0.748,0.820)	(0.919,0.961)	(0.915,0.957)
10	10	3	1	(0.876,0.928)	(0.892,0.940)	(0.754,0.826)	(0.704,0.780)
10	10	3	3	(0.867,0.921)	(0.860,0.916)	(0.860,0.916)	(0.849,0.907)
10	10	3	5	(0.836,0.896)	(0.830,0.890)	(0.898,0.946)	(0.880,0.932)
10	10	5	1	(0.910,0.954)	(0.915,0.957)	(0.763,0.833)	(0.725,0.799)
10	10	5	3	(0.926,0.966)	(0.929,0.967)	(0.860,0.916)	(0.821,0.883)
10	10	5	5	(0.915,0.957)	(0.922,0.962)	(0.898,0.946)	(0.889,0.939)
10	15	1	1	(0.752,0.824)	(0.767,0.837)	(0.748,0.820)	(0.718,0.794)
10	15	1	3	(0.708,0.784)	(0.720,0.796)	(0.860,0.916)	(0.858,0.914)
10	15	1	5	(0.716,0.792)	(0.725,0.799)	(0.919,0.961)	(0.917,0.959)
10	15	3	1	(0.880,0.932)	(0.885,0.935)	(0.739,0.813)	(0.725,0.799)
10	15	3	3	(0.852,0.908)	(0.841,0.899)	(0.865,0.919)	(0.849,0.907)
10	15	3	5	(0.887,0.937)	(0.860,0.916)	(0.915,0.957)	(0.903,0.949)
10	15	5	1	(0.910,0.954)	(0.910,0.954)	(0.723,0.797)	(0.716,0.792)
10	15	5	3	(0.933,0.971)	(0.933,0.971)	(0.876,0.928)	(0.874,0.926)
10	15	5	5	(0.908,0.952)	(0.892,0.940)	(0.919,0.961)	(0.915,0.957)
15	10	1	1	(0.746,0.818)	(0.748,0.820)	(0.739,0.813)	(0.725,0.799)
15	10	1	3	(0.765,0.835)	(0.763,0.833)	(0.849,0.907)	(0.830,0.890)
15	10	1	5	(0.712,0.788)	(0.712,0.788)	(0.896,0.944)	(0.901,0.947)
15	10	3	1	(0.838,0.898)	(0.852,0.908)	(0.712,0.788)	(0.693,0.771)
15	10	3	3	(0.858,0.914)	(0.854,0.910)	(0.889,0.939)	(0.865,0.919)
15	10	3	5	(0.869,0.923)	(0.863,0.917)	(0.876,0.928)	(0.883,0.933)
15	10	5	1	(0.926,0.966)	(0.919,0.961)	(0.729,0.803)	(0.716,0.792)
15	10	5	3	(0.917,0.959)	(0.908,0.952)	(0.841,0.899)	(0.834,0.894)
15	10	5	5	(0.929,0.967)	(0.896,0.944)	(0.922,0.962)	(0.915,0.957)
15	15	1	1	(0.771,0.841)	(0.780,0.848)	(0.704,0.780)	(0.670,0.750)
15	15	1	3	(0.739,0.813)	(0.769,0.839)	(0.843,0.901)	(0.841,0.899)
15	15	1	5	(0.761,0.831)	(0.769,0.839)	(0.917,0.959)	(0.905,0.951)
15	15	3	1	(0.838,0.898)	(0.843,0.901)	(0.725,0.799)	(0.704,0.780)
15	15	3	3	(0.841,0.899)	(0.836,0.896)	(0.841,0.899)	(0.834,0.894)
15	15	3	5	(0.865,0.919)	(0.869,0.923)	(0.922,0.962)	(0.912,0.956)
15	15	5	1	(0.936,0.972)	(0.945,0.979)	(0.714,0.790)	(0.697,0.775)
15	15	5	3	(0.880,0.932)	(0.892,0.940)	(0.863,0.917)	(0.838,0.898)
15	15	5	5	(0.933,0.971)	(0.933,0.971)	(0.919,0.961)	(0.910,0.954)

98

TABLE 1 (continued)

C. $\theta_1 = -0.75$ $\theta_2 = 0.75$

N1	N2	K1	K2	PK1	QK1	PK2	QK2
10	10	1	1	(0.693,0.771)	(0.720,0.796)	(0.710,0.786)	(0.685,0.763)
10	10	1	3	(0.702,0.778)	(0.735,0.809)	(0.827,0.889)	(0.827,0.889)
10	10	1	5	(0.714,0.790)	(0.731,0.805)	(0.915,0.957)	(0.905,0.951)
10	10	3	1	(0.887,0.937)	(0.883,0.933)	(0.720,0.796)	(0.683,0.761)
10	10	3	3	(0.887,0.937)	(0.885,0.935)	(0.832,0.892)	(0.819,0.881)
10	10	3	5	(0.858,0.914)	(0.852,0.908)	(0.915,0.957)	(0.894,0.942)
10	10	5	1	(0.883,0.933)	(0.880,0.932)	(0.718,0.794)	(0.695,0.773)
10	10	5	3	(0.931,0.969)	(0.926,0.966)	(0.845,0.903)	(0.810,0.874)
10	10	5	5	(0.898,0.946)	(0.905,0.951)	(0.903,0.949)	(0.901,0.947)
10	15	1	1	(0.712,0.788)	(0.716,0.792)	(0.752,0.824)	(0.737,0.811)
10	15	1	3	(0.748,0.820)	(0.729,0.803)	(0.880,0.932)	(0.867,0.921)
10	15	1	5	(0.739,0.813)	(0.720,0.796)	(0.929,0.967)	(0.919,0.961)
10	15	3	1	(0.860,0.916)	(0.838,0.898)	(0.708,0.784)	(0.706,0.782)
10	15	3	3	(0.854,0.910)	(0.841,0.899)	(0.871,0.925)	(0.865,0.919)
10	15	3	5	(0.858,0.914)	(0.852,0.908)	(0.910,0.954)	(0.915,0.957)
10	15	5	1	(0.931,0.969)	(0.929,0.967)	(0.752,0.824)	(0.731,0.805)
10	15	5	3	(0.912,0.956)	(0.903,0.949)	(0.874,0.926)	(0.860,0.916)
10	15	5	5	(0.926,0.966)	(0.917,0.959)	(0.926,0.966)	(0.910,0.954)
15	10	1	1	(0.691,0.769)	(0.689,0.767)	(0.769,0.839)	(0.752,0.824)
15	10	1	3	(0.729,0.803)	(0.714,0.790)	(0.854,0.910)	(0.838,0.898)
15	10	1	5	(0.716,0.792)	(0.725,0.799)	(0.892,0.940)	(0.892,0.940)
15	10	3	1	(0.876,0.928)	(0.880,0.932)	(0.742,0.814)	(0.737,0.811)
15	10	3	3	(0.863,0.917)	(0.847,0.905)	(0.878,0.930)	(0.874,0.926)
15	10	3	5	(0.852,0.908)	(0.841,0.899)	(0.908,0.952)	(0.901,0.947)
15	10	5	1	(0.917,0.959)	(0.910,0.954)	(0.731,0.805)	(0.718,0.794)
15	10	5	3	(0.908,0.952)	(0.912,0.956)	(0.849,0.907)	(0.847,0.905)
15	10	5	5	(0.938,0.974)	(0.912,0.956)	(0.896,0.944)	(0.878,0.930)
15	15	1	1	(0.737,0.811)	(0.759,0.829)	(0.737,0.811)	(0.712,0.788)
15	15	1	3	(0.750,0.822)	(0.773,0.843)	(0.878,0.930)	(0.843,0.901)
15	15	1	5	(0.765,0.835)	(0.791,0.857)	(0.931,0.969)	(0.922,0.962)
15	15	3	1	(0.863,0.917)	(0.865,0.919)	(0.752,0.824)	(0.720,0.796)
15	15	3	3	(0.863,0.917)	(0.856,0.912)	(0.863,0.917)	(0.841,0.899)
15	15	3	5	(0.860,0.916)	(0.863,0.917)	(0.931,0.969)	(0.910,0.954)
15	15	5	1	(0.931,0.969)	(0.926,0.966)	(0.689,0.767)	(0.672,0.752)
15	15	5	3	(0.926,0.966)	(0.929,0.967)	(0.834,0.894)	(0.830,0.890)
15	15	5	5	(0.922,0.962)	(0.922,0.962)	(0.926,0.966)	(0.922,0.962)

99

D. $\theta_1 = -0.5$ $\theta_2 = 0.5$

N1	N2	K1	K2	PK1	QK1	PK2	QK2
10	10	1	1	(0.596,0.680)	(0.606,0.690)	(0.625,0.707)	(0.602,0.686)
10	10	1	3	(0.627,0.709)	(0.652,0.732)	(0.725,0.799)	(0.710,0.786)
10	10	1	5	(0.616,0.700)	(0.639,0.721)	(0.806,0.870)	(0.801,0.867)
10	10	3	1	(0.767,0.837)	(0.756,0.828)	(0.621,0.703)	(0.594,0.678)
10	10	3	3	(0.791,0.857)	(0.797,0.863)	(0.710,0.786)	(0.685,0.763)
10	10	3	5	(0.720,0.796)	(0.737,0.811)	(0.786,0.854)	(0.767,0.837)
10	10	5	1	(0.776,0.844)	(0.769,0.839)	(0.643,0.725)	(0.618,0.702)
10	10	5	3	(0.817,0.879)	(0.819,0.881)	(0.695,0.773)	(0.676,0.756)
10	10	5	5	(0.782,0.850)	(0.791,0.857)	(0.799,0.865)	(0.763,0.833)
10	15	1	1	(0.631,0.713)	(0.645,0.727)	(0.633,0.715)	(0.647,0.729)
10	15	1	3	(0.649,0.731)	(0.658,0.738)	(0.782,0.850)	(0.786,0.854)
10	15	1	5	(0.639,0.721)	(0.652,0.732)	(0.843,0.901)	(0.821,0.883)
10	15	3	1	(0.735,0.809)	(0.737,0.811)	(0.618,0.702)	(0.598,0.682)
10	15	3	3	(0.761,0.831)	(0.744,0.816)	(0.750,0.822)	(0.731,0.805)
10	15	3	5	(0.761,0.831)	(0.744,0.816)	(0.801,0.867)	(0.784,0.852)
10	15	5	1	(0.845,0.903)	(0.836,0.896)	(0.658,0.738)	(0.654,0.734)
10	15	5	3	(0.812,0.876)	(0.795,0.861)	(0.737,0.811)	(0.733,0.807)
10	15	5	5	(0.804,0.868)	(0.804,0.868)	(0.834,0.894)	(0.808,0.872)
15	10	1	1	(0.621,0.703)	(0.612,0.696)	(0.689,0.767)	(0.660,0.740)
15	10	1	3	(0.629,0.711)	(0.631,0.713)	(0.735,0.809)	(0.729,0.803)
15	10	1	5	(0.656,0.736)	(0.660,0.740)	(0.808,0.872)	(0.795,0.861)
15	10	3	1	(0.765,0.835)	(0.759,0.829)	(0.664,0.744)	(0.658,0.738)
15	10	3	3	(0.731,0.805)	(0.742,0.814)	(0.750,0.822)	(0.723,0.797)
15	10	3	5	(0.756,0.828)	(0.744,0.816)	(0.795,0.861)	(0.773,0.843)
15	10	5	1	(0.799,0.865)	(0.795,0.861)	(0.643,0.725)	(0.641,0.723)
15	10	5	3	(0.808,0.872)	(0.797,0.863)	(0.752,0.824)	(0.733,0.807)
15	10	5	5	(0.821,0.883)	(0.819,0.881)	(0.797,0.863)	(0.791,0.857)
15	15	1	1	(0.645,0.727)	(0.660,0.740)	(0.639,0.721)	(0.623,0.705)
15	15	1	3	(0.658,0.738)	(0.670,0.750)	(0.754,0.826)	(0.718,0.794)
15	15	1	5	(0.666,0.746)	(0.697,0.775)	(0.836,0.896)	(0.832,0.892)
15	15	3	1	(0.756,0.828)	(0.756,0.828)	(0.639,0.721)	(0.623,0.705)
15	15	3	3	(0.742,0.814)	(0.733,0.807)	(0.729,0.803)	(0.714,0.790)
15	15	3	5	(0.782,0.850)	(0.778,0.846)	(0.819,0.881)	(0.799,0.865)
15	15	5	1	(0.821,0.883)	(0.827,0.889)	(0.610,0.694)	(0.608,0.692)
15	15	5	3	(0.836,0.896)	(0.823,0.885)	(0.735,0.809)	(0.716,0.792)
15	15	5	5	(0.841,0.899)	(0.825,0.887)	(0.836,0.896)	(0.810,0.874)

TABLE 1 (continued)

E. $\theta_1 = -0.3$ $\theta_2 = 0.2$

N1	N2	K1	K2	PK1	QK1	PK2	QK2
10	10	1	1	(0.531,0.617)	(0.547,0.633)	(0.533,0.619)	(0.516,0.604)
10	10	1	3	(0.523,0.609)	(0.537,0.623)	(0.592,0.676)	(0.553,0.639)
10	10	1	5	(0.512,0.600)	(0.508,0.596)	(0.625,0.707)	(0.602,0.686)
10	10	3	1	(0.588,0.672)	(0.588,0.672)	(0.537,0.623)	(0.508,0.596)
10	10	3	3	(0.567,0.653)	(0.602,0.686)	(0.579,0.665)	(0.565,0.651)
10	10	3	5	(0.502,0.590)	(0.527,0.613)	(0.604,0.688)	(0.594,0.678)
10	10	5	1	(0.582,0.666)	(0.602,0.686)	(0.563,0.649)	(0.537,0.623)
10	10	5	3	(0.590,0.674)	(0.621,0.703)	(0.567,0.653)	(0.545,0.631)
10	10	5	5	(0.618,0.702)	(0.600,0.684)	(0.598,0.682)	(0.590,0.674)
10	15	1	1	(0.531,0.617)	(0.514,0.602)	(0.539,0.625)	(0.541,0.627)
10	15	1	3	(0.490,0.578)	(0.500,0.588)	(0.606,0.690)	(0.623,0.705)
10	15	1	5	(0.521,0.607)	(0.541,0.627)	(0.625,0.707)	(0.606,0.690)
10	15	3	1	(0.598,0.682)	(0.586,0.670)	(0.533,0.619)	(0.529,0.615)
10	15	3	3	(0.577,0.663)	(0.588,0.672)	(0.600,0.684)	(0.590,0.674)
10	15	3	5	(0.590,0.674)	(0.563,0.649)	(0.608,0.692)	(0.625,0.707)
10	15	5	1	(0.575,0.661)	(0.569,0.655)	(0.523,0.609)	(0.525,0.611)
10	15	5	3	(0.608,0.692)	(0.600,0.684)	(0.618,0.702)	(0.612,0.696)
10	15	5	5	(0.592,0.676)	(0.588,0.672)	(0.582,0.666)	(0.590,0.674)
15	10	1	1	(0.582,0.666)	(0.575,0.661)	(0.510,0.598)	(0.494,0.582)
15	10	1	3	(0.535,0.621)	(0.551,0.637)	(0.547,0.633)	(0.541,0.627)
15	10	1	5	(0.539,0.625)	(0.541,0.627)	(0.569,0.655)	(0.565,0.651)
15	10	3	1	(0.571,0.657)	(0.602,0.686)	(0.519,0.605)	(0.514,0.602)
15	10	3	3	(0.586,0.670)	(0.598,0.682)	(0.600,0.684)	(0.586,0.670)
15	10	3	5	(0.590,0.674)	(0.608,0.692)	(0.592,0.676)	(0.579,0.665)
15	10	5	1	(0.604,0.688)	(0.621,0.703)	(0.525,0.611)	(0.519,0.605)
15	10	5	3	(0.610,0.694)	(0.623,0.705)	(0.563,0.649)	(0.555,0.641)
15	10	5	5	(0.625,0.707)	(0.627,0.709)	(0.586,0.670)	(0.592,0.676)
15	15	1	1	(0.569,0.655)	(0.565,0.651)	(0.504,0.592)	(0.482,0.570)
15	15	1	3	(0.523,0.609)	(0.549,0.635)	(0.565,0.651)	(0.537,0.623)
15	15	1	5	(0.531,0.617)	(0.561,0.647)	(0.658,0.738)	(0.618,0.702)
15	15	3	1	(0.557,0.643)	(0.563,0.649)	(0.541,0.627)	(0.502,0.590)
15	15	3	3	(0.608,0.692)	(0.584,0.668)	(0.588,0.672)	(0.563,0.649)
15	15	3	5	(0.588,0.672)	(0.592,0.676)	(0.602,0.686)	(0.598,0.682)
15	15	5	1	(0.635,0.717)	(0.645,0.727)	(0.553,0.639)	(0.547,0.633)
15	15	5	3	(0.612,0.696)	(0.602,0.686)	(0.561,0.647)	(0.541,0.627)
15	15	5	5	(0.633,0.715)	(0.652,0.732)	(0.641,0.723)	(0.623,0.705)

101

TABLE 2   95% Confidence Intervals for the Probabilities of Correct Classification

(Logistic Distribution with Mean θ and Variance 1)

A. $\theta_1 = -1$ $\theta_2 = 1$

N1	N2	K1	K2	PK1	QK1	PK2	QK2
10	10	1	1	(0.825,0.887)	(0.847,0.905)	(0.825,0.887)	(0.786,0.854)
10	10	1	3	(0.819,0.881)	(0.838,0.898)	(0.931,0.969)	(0.912,0.956)
10	10	1	5	(0.830,0.890)	(0.843,0.901)	(0.970,0.994)	(0.950,0.982)
10	10	3	1	(0.926,0.966)	(0.943,0.977)	(0.817,0.879)	(0.788,0.856)
10	10	3	3	(0.955,0.985)	(0.950,0.982)	(0.915,0.957)	(0.912,0.956)
10	10	3	5	(0.922,0.962)	(0.929,0.967)	(0.958,0.986)	(0.950,0.982)
10	10	5	1	(0.960,0.988)	(0.965,0.991)	(0.791,0.857)	(0.746,0.818)
10	10	5	3	(0.963,0.989)	(0.958,0.986)	(0.931,0.969)	(0.915,0.957)
10	10	5	5	(0.973,0.995)	(0.970,0.994)	(0.960,0.988)	(0.948,0.980)
10	15	1	1	(0.797,0.863)	(0.814,0.878)	(0.858,0.914)	(0.854,0.910)
10	15	1	3	(0.817,0.879)	(0.825,0.887)	(0.936,0.972)	(0.926,0.966)
10	15	1	5	(0.825,0.887)	(0.827,0.889)	(0.955,0.985)	(0.955,0.985)
10	15	3	1	(0.936,0.972)	(0.938,0.974)	(0.830,0.890)	(0.786,0.854)
10	15	3	3	(0.929,0.967)	(0.931,0.969)	(0.926,0.966)	(0.915,0.957)
10	15	3	5	(0.940,0.976)	(0.936,0.972)	(0.960,0.988)	(0.948,0.980)
10	15	5	1	(0.960,0.988)	(0.958,0.986)	(0.821,0.883)	(0.812,0.876)
10	15	5	3	(0.968,0.992)	(0.968,0.992)	(0.948,0.980)	(0.926,0.966)
10	15	5	5	(0.968,0.992)	(0.960,0.988)	(0.968,0.992)	(0.965,0.991)
15	10	1	1	(0.867,0.921)	(0.858,0.914)	(0.821,0.883)	(0.799,0.865)
15	10	1	3	(0.791,0.857)	(0.797,0.863)	(0.924,0.964)	(0.915,0.957)
15	10	1	5	(0.799,0.865)	(0.817,0.879)	(0.960,0.988)	(0.950,0.982)
15	10	3	1	(0.929,0.967)	(0.917,0.959)	(0.810,0.874)	(0.795,0.861)
15	10	3	3	(0.940,0.976)	(0.938,0.974)	(0.922,0.962)	(0.919,0.961)
15	10	3	5	(0.938,0.974)	(0.922,0.962)	(0.968,0.992)	(0.965,0.991)
15	10	5	1	(0.968,0.992)	(0.968,0.992)	(0.825,0.887)	(0.801,0.867)
15	10	5	3	(0.960,0.988)	(0.958,0.986)	(0.926,0.966)	(0.915,0.957)
15	10	5	5	(0.963,0.989)	(0.955,0.985)	(0.970,0.994)	(0.973,0.995)
15	15	1	1	(0.814,0.878)	(0.834,0.894)	(0.795,0.861)	(0.776,0.844)
15	15	1	3	(0.806,0.870)	(0.814,0.878)	(0.938,0.974)	(0.933,0.971)
15	15	1	5	(0.797,0.863)	(0.812,0.876)	(0.973,0.995)	(0.965,0.991)
15	15	3	1	(0.926,0.966)	(0.922,0.962)	(0.823,0.885)	(0.812,0.876)
15	15	3	3	(0.950,0.982)	(0.938,0.974)	(0.936,0.972)	(0.929,0.967)
15	15	3	5	(0.931,0.969)	(0.931,0.969)	(0.958,0.986)	(0.958,0.986)
15	15	5	1	(0.965,0.991)	(0.963,0.989)	(0.821,0.883)	(0.801,0.867)
15	15	5	3	(0.958,0.986)	(0.960,0.988)	(0.924,0.964)	(0.924,0.964)
15	15	5	5	(0.968,0.992)	(0.960,0.988)	(0.963,0.989)	(0.955,0.985)

TABLE 2 (continued)

B. $\theta_1 = -1$ $\theta_2 = 0.5$

N1	N2	K1	K2	PK1	QK1	PK2	QK2
10	10	1	1	(0.729,0.803)	(0.761,0.831)	(0.748,0.820)	(0.697,0.775)
10	10	1	3	(0.748,0.820)	(0.778,0.846)	(0.885,0.935)	(0.843,0.901)
10	10	1	5	(0.746,0.818)	(0.771,0.841)	(0.924,0.964)	(0.917,0.959)
10	10	3	1	(0.863,0.917)	(0.878,0.930)	(0.742,0.814)	(0.702,0.778)
10	10	3	3	(0.874,0.926)	(0.880,0.932)	(0.869,0.923)	(0.852,0.908)
10	10	3	5	(0.849,0.907)	(0.869,0.923)	(0.903,0.949)	(0.874,0.926)
10	10	5	1	(0.926,0.966)	(0.933,0.971)	(0.771,0.841)	(0.746,0.818)
10	10	5	3	(0.901,0.947)	(0.915,0.957)	(0.876,0.928)	(0.852,0.908)
10	10	5	5	(0.924,0.964)	(0.908,0.952)	(0.922,0.962)	(0.901,0.947)
10	15	1	1	(0.748,0.820)	(0.744,0.816)	(0.731,0.805)	(0.699,0.777)
10	15	1	3	(0.759,0.829)	(0.773,0.843)	(0.878,0.930)	(0.871,0.925)
10	15	1	5	(0.763,0.833)	(0.763,0.833)	(0.926,0.966)	(0.901,0.947)
10	15	3	1	(0.843,0.901)	(0.849,0.907)	(0.786,0.854)	(0.761,0.831)
10	15	3	3	(0.887,0.937)	(0.885,0.935)	(0.883,0.933)	(0.858,0.914)
10	15	3	5	(0.894,0.942)	(0.896,0.944)	(0.938,0.974)	(0.929,0.967)
10	15	5	1	(0.903,0.949)	(0.910,0.954)	(0.780,0.848)	(0.764,0.818)
10	15	5	3	(0.917,0.959)	(0.919,0.961)	(0.858,0.914)	(0.845,0.903)
10	15	5	5	(0.917,0.959)	(0.929,0.967)	(0.931,0.969)	(0.915,0.957)
15	10	1	1	(0.782,0.850)	(0.791,0.857)	(0.773,0.843)	(0.737,0.811)
15	10	1	3	(0.750,0.822)	(0.750,0.822)	(0.865,0.919)	(0.858,0.914)
15	10	1	5	(0.733,0.807)	(0.735,0.809)	(0.926,0.966)	(0.912,0.956)
15	10	3	1	(0.858,0.914)	(0.845,0.903)	(0.759,0.829)	(0.729,0.803)
15	10	3	3	(0.863,0.917)	(0.867,0.921)	(0.845,0.903)	(0.843,0.901)
15	10	3	5	(0.889,0.939)	(0.878,0.930)	(0.943,0.977)	(0.931,0.969)
15	10	5	1	(0.917,0.959)	(0.919,0.961)	(0.742,0.814)	(0.729,0.803)
15	10	5	3	(0.910,0.954)	(0.912,0.956)	(0.867,0.921)	(0.849,0.907)
15	10	5	5	(0.912,0.956)	(0.908,0.952)	(0.912,0.956)	(0.892,0.940)
15	15	1	1	(0.750,0.822)	(0.767,0.837)	(0.778,0.846)	(0.754,0.826)
15	15	1	3	(0.795,0.861)	(0.801,0.867)	(0.878,0.930)	(0.863,0.917)
15	15	1	5	(0.744,0.816)	(0.767,0.837)	(0.929,0.967)	(0.915,0.957)
15	15	3	1	(0.871,0.925)	(0.885,0.935)	(0.735,0.809)	(0.704,0.780)
15	15	3	3	(0.867,0.921)	(0.883,0.933)	(0.869,0.923)	(0.856,0.912)
15	15	3	5	(0.876,0.928)	(0.903,0.949)	(0.929,0.967)	(0.917,0.959)
15	15	5	1	(0.919,0.961)	(0.917,0.959)	(0.759,0.829)	(0.729,0.803)
15	15	5	3	(0.917,0.959)	(0.922,0.962)	(0.867,0.921)	(0.874,0.926)
15	15	5	5	(0.924,0.964)	(0.926,0.966)	(0.933,0.971)	(0.926,0.966)

TABLE 2 (continued)

C. $\theta_1 = -0.75$ $\theta_2 = 0.75$

N1	N2	K1	K2	PK1	QK1	PK2	QK2
10	10	1	1	(0.769,0.839)	(0.791,0.857)	(0.769,0.839)	(0.714,0.790)
10	10	1	3	(0.771,0.841)	(0.778,0.846)	(0.860,0.916)	(0.860,0.916)
10	10	1	5	(0.729,0.803)	(0.742,0.814)	(0.910,0.954)	(0.905,0.951)
10	10	3	1	(0.867,0.921)	(0.885,0.935)	(0.742,0.814)	(0.725,0.799)
10	10	3	3	(0.852,0.908)	(0.747,0.905)	(0.871,0.925)	(0.854,0.910)
10	10	3	5	(0.845,0.912)	(0.856,0.912)	(0.901,0.947)	(0.908,0.952)
10	10	5	1	(0.915,0.957)	(0.915,0.957)	(0.808,0.872)	(0.767,0.837)
10	10	5	3	(0.929,0.967)	(0.924,0.964)	(0.858,0.914)	(0.854,0.910)
10	10	5	5	(0.917,0.959)	(0.922,0.962)	(0.905,0.951)	(0.898,0.946)
10	15	1	1	(0.729,0.803)	(0.735,0.809)	(0.739,0.813)	(0.712,0.788)
10	15	1	3	(0.733,0.807)	(0.742,0.814)	(0.880,0.932)	(0.878,0.930)
10	15	1	5	(0.759,0.829)	(0.763,0.833)	(0.917,0.959)	(0.908,0.952)
10	15	3	1	(0.876,0.928)	(0.867,0.921)	(0.750,0.822)	(0.733,0.807)
10	15	3	3	(0.865,0.919)	(0.867,0.921)	(0.845,0.903)	(0.838,0.898)
10	15	3	5	(0.821,0.883)	(0.836,0.896)	(0.936,0.972)	(0.924,0.964)
10	15	5	1	(0.915,0.957)	(0.908,0.952)	(0.737,0.811)	(0.756,0.828)
10	15	5	3	(0.905,0.951)	(0.919,0.961)	(0.867,0.921)	(0.854,0.910)
10	15	5	5	(0.898,0.946)	(0.898,0.946)	(0.931,0.969)	(0.915,0.957)
15	10	1	1	(0.733,0.807)	(0.739,0.813)	(0.744,0.816)	(0.733,0.807)
15	10	1	3	(0.742,0.814)	(0.748,0.820)	(0.880,0.932)	(0.871,0.925)
15	10	1	5	(0.793,0.859)	(0.795,0.861)	(0.929,0.967)	(0.931,0.969)
15	10	3	1	(0.876,0.928)	(0.880,0.932)	(0.742,0.814)	(0.737,0.811)
15	10	3	3	(0.878,0.930)	(0.871,0.925)	(0.878,0.930)	(0.867,0.921)
15	10	3	5	(0.871,0.925)	(0.856,0.912)	(0.926,0.966)	(0.931,0.969)
15	10	5	1	(0.910,0.954)	(0.915,0.957)	(0.742,0.814)	(0.733,0.807)
15	10	5	3	(0.938,0.974)	(0.924,0.964)	(0.880,0.932)	(0.863,0.917)
15	10	5	5	(0.894,0.942)	(0.901,0.947)	(0.922,0.962)	(0.924,0.964)
15	15	1	1	(0.765,0.835)	(0.776,0.844)	(0.731,0.805)	(0.712,0.788)
15	15	1	3	(0.761,0.831)	(0.780,0.848)	(0.887,0.937)	(0.876,0.928)
15	15	1	5	(0.769,0.839)	(0.788,0.856)	(0.905,0.951)	(0.915,0.957)
15	15	3	1	(0.856,0.912)	(0.849,0.907)	(0.769,0.839)	(0.744,0.816)
15	15	3	3	(0.869,0.923)	(0.871,0.925)	(0.874,0.926)	(0.869,0.923)
15	15	3	5	(0.887,0.937)	(0.880,0.932)	(0.919,0.961)	(0.912,0.956)
15	15	5	1	(0.958,0.986)	(0.955,0.985)	(0.769,0.839)	(0.739,0.813)
15	15	5	3	(0.936,0.972)	(0.938,0.974)	(0.883,0.933)	(0.867,0.921)
15	15	5	5	(0.910,0.954)	(0.908,0.952)	(0.950,0.982)	(0.945,0.979)

TABLE 2 (continued)

D. $\theta_1 = -0.5$ $\theta_2 = 0.5$

N1	N2	K1	K2	PK1	QK1	PK2	QK2
10	10	1	1	(0.670,0.750)	(0.674,0.754)	(0.664,0.744)	(0.618,0.702)
10	10	1	3	(0.668,0.748)	(0.689,0.767)	(0.761,0.831)	(0.748,0.820)
10	10	1	5	(0.652,0.732)	(0.668,0.748)	(0.806,0.870)	(0.810,0.874)
10	10	3	1	(0.735,0.809)	(0.756,0.828)	(0.654,0.734)	(0.604,0.688)
10	10	3	3	(0.763,0.833)	(0.763,0.833)	(0.746,0.818)	(0.737,0.811)
10	10	3	5	(0.795,0.861)	(0.810,0.874)	(0.801,0.867)	(0.799,0.865)
10	10	5	1	(0.788,0.856)	(0.801,0.867)	(0.668,0.748)	(0.629,0.711)
10	10	5	3	(0.767,0.837)	(0.769,0.839)	(0.754,0.826)	(0.737,0.811)
10	10	5	5	(0.819,0.881)	(0.832,0.892)	(0.810,0.874)	(0.797,0.863)
10	15	1	1	(0.656,0.736)	(0.660,0.740)	(0.702,0.778)	(0.679,0.757)
10	15	1	3	(0.670,0.750)	(0.693,0.771)	(0.776,0.844)	(0.763,0.833)
10	15	1	5	(0.656,0.736)	(0.654,0.734)	(0.817,0.879)	(0.791,0.857)
10	15	3	1	(0.767,0.837)	(0.782,0.850)	(0.683,0.761)	(0.685,0.763)
10	15	3	3	(0.723,0.797)	(0.739,0.813)	(0.761,0.831)	(0.759,0.829)
10	15	3	5	(0.714,0.790)	(0.716,0.792)	(0.849,0.907)	(0.819,0.881)
10	15	5	1	(0.804,0.868)	(0.819,0.881)	(0.681,0.759)	(0.676,0.756)
10	15	5	3	(0.812,0.876)	(0.823,0.885)	(0.771,0.841)	(0.776,0.844)
10	15	5	5	(0.838,0.898)	(0.838,0.898)	(0.801,0.867)	(0.804,0.868)
15	10	1	1	(0.681,0.759)	(0.706,0.782)	(0.693,0.771)	(0.664,0.744)
15	10	1	3	(0.641,0.723)	(0.662,0.742)	(0.748,0.820)	(0.739,0.813)
15	10	1	5	(0.662,0.742)	(0.656,0.736)	(0.812,0.876)	(0.804,0.868)
15	10	3	1	(0.750,0.822)	(0.739,0.813)	(0.654,0.734)	(0.658,0.738)
15	10	3	3	(0.780,0.848)	(0.778,0.846)	(0.767,0.837)	(0.771,0.841)
15	10	3	5	(0.769,0.839)	(0.771,0.841)	(0.817,0.879)	(0.814,0.878)
15	10	5	1	(0.834,0.894)	(0.832,0.892)	(0.656,0.736)	(0.649,0.731)
15	10	5	3	(0.814,0.878)	(0.823,0.885)	(0.776,0.844)	(0.754,0.826)
15	10	5	5	(0.841,0.899)	(0.845,0.903)	(0.830,0.890)	(0.823,0.885)
15	15	1	1	(0.658,0.738)	(0.668,0.748)	(0.672,0.752)	(0.672,0.752)
15	15	1	3	(0.631,0.713)	(0.660,0.740)	(0.776,0.844)	(0.761,0.831)
15	15	1	5	(0.654,0.734)	(0.660,0.740)	(0.834,0.894)	(0.838,0.898)
15	15	3	1	(0.744,0.816)	(0.748,0.820)	(0.689,0.767)	(0.668,0.748)
15	15	3	3	(0.765,0.835)	(0.759,0.829)	(0.776,0.844)	(0.769,0.839)
15	15	3	5	(0.765,0.835)	(0.797,0.863)	(0.825,0.887)	(0.814,0.787)
15	15	5	1	(0.849,0.907)	(0.858,0.914)	(0.681,0.759)	(0.647,0.729)
15	15	5	3	(0.843,0.901)	(0.834,0.894)	(0.771,0.841)	(0.763,0.833)
15	15	5	5	(0.852,0.908)	(0.847,0.905)	(0.776,0.844)	(0.788,0.856)

105

TABLE 2 (continued)

E. $\theta_1 = -0.3$ $\theta_2 = 0.2$

N1	N2	K1	K2	PK1	QK1	PK2	QK2
10	10	1	1	(0.545,0.631)	(0.594,0.678)	(0.569,0.655)	(0.531,0.617)
10	10	1	3	(0.553,0.639)	(0.569,0.655)	(0.577,0.663)	(0.537,0.623)
10	10	1	5	(0.535,0.621)	(0.569,0.655)	(0.608,0.692)	(0.577,0.663)
10	10	3	1	(0.594,0.678)	(0.602,0.686)	(0.525,0.611)	(0.508,0.596)
10	10	3	3	(0.561,0.647)	(0.604,0.688)	(0.598,0.682)	(0.579,0.665)
10	10	3	5	(0.549,0.635)	(0.563,0.649)	(0.610,0.694)	(0.598,0.682)
10	10	5	1	(0.571,0.657)	(0.594,0.678)	(0.523,0.609)	(0.510,0.598)
10	10	5	3	(0.614,0.698)	(0.616,0.700)	(0.573,0.659)	(0.545,0.631)
10	10	5	5	(0.618,0.702)	(0.633,0.715)	(0.600,0.684)	(0.606,0.690)
10	15	1	1	(0.496,0.584)	(0.521,0.607)	(0.543,0.629)	(0.555,0.641)
10	15	1	3	(0.539,0.625)	(0.539,0.625)	(0.592,0.676)	(0.573,0.659)
10	15	1	5	(0.514,0.602)	(0.519,0.605)	(0.629,0.711)	(0.625,0.707)
10	15	3	1	(0.633,0.715)	(0.637,0.719)	(0.514,0.602)	(0.492,0.580)
10	15	3	3	(0.577,0.663)	(0.565,0.651)	(0.614,0.698)	(0.614,0.698)
10	15	3	5	(0.594,0.678)	(0.590,0.674)	(0.645,0.727)	(0.645,0.727)
10	15	5	1	(0.627,0.709)	(0.637,0.719)	(0.557,0.643)	(0.543,0.629)
10	15	5	3	(0.594,0.678)	(0.610,0.694)	(0.600,0.684)	(0.600,0.684)
10	15	5	5	(0.565,0.651)	(0.579,0.665)	(0.633,0.715)	(0.652,0.732)
15	10	1	1	(0.573,0.659)	(0.592,0.676)	(0.531,0.617)	(0.512,0.600)
15	10	1	3	(0.525,0.611)	(0.514,0.602)	(0.582,0.666)	(0.582,0.666)
15	10	1	5	(0.563,0.649)	(0.569,0.655)	(0.618,0.702)	(0.627,0.709)
15	10	3	1	(0.623,0.705)	(0.610,0.694)	(0.567,0.653)	(0.551,0.637)
15	10	3	3	(0.616,0.700)	(0.596,0.680)	(0.555,0.641)	(0.559,0.645)
15	10	3	5	(0.592,0.676)	(0.592,0.676)	(0.590,0.674)	(0.586,0.670)
15	10	5	1	(0.664,0.744)	(0.670,0.750)	(0.533,0.619)	(0.527,0.613)
15	10	5	3	(0.637,0.719)	(0.641,0.723)	(0.579,0.665)	(0.586,0.670)
15	10	5	5	(0.654,0.734)	(0.658,0.738)	(0.594,0.678)	(0.608,0.692)
15	15	1	1	(0.527,0.613)	(0.551,0.637)	(0.543,0.629)	(0.527,0.613)
15	15	1	3	(0.553,0.639)	(0.561,0.647)	(0.635,0.717)	(0.610,0.694)
15	15	1	5	(0.551,0.637)	(0.582,0.666)	(0.625,0.707)	(0.635,0.717)
15	15	3	1	(0.596,0.680)	(0.610,0.694)	(0.586,0.670)	(0.577,0.663)
15	15	3	3	(0.551,0.637)	(0.571,0.657)	(0.573,0.659)	(0.551,0.637)
15	15	3	5	(0.582,0.666)	(0.606,0.690)	(0.567,0.653)	(0.573,0.659)
15	15	5	1	(0.604,0.688)	(0.596,0.680)	(0.586,0.670)	(0.563,0.649)
15	15	5	3	(0.668,0.748)	(0.664,0.744)	(0.631,0.713)	(0.602,0.686)
15	15	5	5	(0.623,0.705)	(0.637,0.719)	(0.643,0.725)	(0.618,0.702)

106

TABLE 3   95% Confidence Intervals for the Probabilities of Correct Classification

(Cauchy Distribution With Location Parameter θ and Scale Parameter 1)

A. $\theta_1=-1$ $\theta_2=1$

N1	N2	K1	K2	PK1	QK1	PK2	QK2
10	10	1	1	(0.691,0.769)	(0.699,0.777)	(0.691,0.769)	(0.647,0.729)
10	10	1	3	(0.676,0.756)	(0.676,0.756)	(0.784,0.852)	(0.708,0.784)
10	10	1	5	(0.687,0.765)	(0.687,0.765)	(0.845,0.903)	(0.791,0.857)
10	10	3	1	(0.799,0.865)	(0.765,0.835)	(0.693,0.771)	(0.679,0.757)
10	10	3	3	(0.771,0.841)	(0.756,0.828)	(0.797,0.863)	(0.718,0.794)
10	10	3	5	(0.817,0.879)	(0.769,0.839)	(0.830,0.890)	(0.773,0.843)
10	10	5	1	(0.852,0.908)	(0.791,0.857)	(0.723,0.797)	(0.674,0.754)
10	10	5	3	(0.823,0.885)	(0.767,0.837)	(0.748,0.820)	(0.697,0.775)
10	10	5	5	(0.832,0.892)	(0.784,0.852)	(0.847,0.905)	(0.771,0.841)
10	15	1	1	(0.704,0.780)	(0.710,0.786)	(0.739,0.813)	(0.699,0.777)
10	15	1	3	(0.704,0.780)	(0.712,0.788)	(0.782,0.850)	(0.716,0.792)
10	15	1	5	(0.710,0.786)	(0.699,0.777)	(0.845,0.903)	(0.771,0.841)
10	15	3	1	(0.797,0.863)	(0.733,0.807)	(0.746,0.818)	(0.723,0.797)
10	15	3	3	(0.776,0.844)	(0.761,0.831)	(0.810,0.874)	(0.744,0.816)
10	15	3	5	(0.788,0.856)	(0.735,0.809)	(0.860,0.916)	(0.778,0.846)
10	15	5	1	(0.836,0.896)	(0.799,0.865)	(0.720,0.796)	(0.706,0.782)
10	15	5	3	(0.856,0.912)	(0.782,0.850)	(0.812,0.876)	(0.739,0.813)
10	15	5	5	(0.854,0.910)	(0.769,0.839)	(0.854,0.910)	(0.763,0.833)
15	10	1	1	(0.693,0.771)	(0.691,0.769)	(0.720,0.796)	(0.712,0.788)
15	10	1	3	(0.720,0.796)	(0.712,0.788)	(0.771,0.841)	(0.746,0.818)
15	10	1	5	(0.683,0.761)	(0.679,0.757)	(0.834,0.894)	(0.780,0.848)
15	10	3	1	(0.808,0.872)	(0.742,0.814)	(0.699,0.777)	(0.679,0.757)
15	10	3	3	(0.806,0.870)	(0.735,0.809)	(0.759,0.829)	(0.689,0.767)
15	10	3	5	(0.827,0.889)	(0.759,0.829)	(0.845,0.903)	(0.797,0.863)
15	10	5	1	(0.867,0.921)	(0.804,0.868)	(0.691,0.769)	(0.685,0.763)
15	10	5	3	(0.880,0.932)	(0.797,0.863)	(0.797,0.863)	(0.725,0.799)
15	10	5	5	(0.843,0.901)	(0.776,0.844)	(0.834,0.894)	(0.804,0.868)
15	15	1	1	(0.723,0.797)	(0.729,0.803)	(0.691,0.769)	(0.652,0.732)
15	15	1	3	(0.695,0.773)	(0.714,0.790)	(0.797,0.863)	(0.723,0.797)
15	15	1	5	(0.704,0.780)	(0.693,0.771)	(0.845,0.903)	(0.776,0.844)
15	15	3	1	(0.819,0.881)	(0.765,0.835)	(0.725,0.799)	(0.687,0.765)
15	15	3	3	(0.788,0.856)	(0.763,0.833)	(0.799,0.865)	(0.710,0.786)
15	15	3	5	(0.821,0.883)	(0.767,0.837)	(0.863,0.917)	(0.773,0.843)
15	15	5	1	(0.849,0.907)	(0.795,0.861)	(0.685,0.763)	(0.672,0.752)
15	15	5	3	(0.863,0.917)	(0.806,0.870)	(0.812,0.876)	(0.733,0.807)
15	15	5	5	(0.869,0.923)	(0.793,0.859)	(0.867,0.921)	(0.784,0.852)

TABLE 3 (continued)

B. $\theta_1 = -1$ $\theta_2 = 0.5$

N1	N2	K1	K2	PK1	QK1	PK2	QK2
10	10	1	1	(0.631,0.713)	(0.658,0.738)	(0.612,0.696)	(0.590,0.674)
10	10	1	3	(0.641,0.723)	(0.660,0.740)	(0.679,0.757)	(0.606,0.690)
10	10	1	5	(0.652,0.732)	(0.639,0.721)	(0.799,0.865)	(0.666,0.746)
10	10	3	1	(0.712,0.788)	(0.689,0.767)	(0.656,0.736)	(0.618,0.702)
10	10	3	3	(0.712,0.788)	(0.691,0.769)	(0.735,0.809)	(0.670,0.750)
10	10	3	5	(0.739,0.813)	(0.683,0.761)	(0.750,0.822)	(0.689,0.767)
10	10	5	1	(0.756,0.828)	(0.735,0.809)	(0.654,0.734)	(0.635,0.717)
10	10	5	3	(0.786,0.854)	(0.756,0.828)	(0.765,0.835)	(0.737,0.811)
10	10	5	5	(0.759,0.829)	(0.718,0.794)	(0.780,0.848)	(0.704,0.780)
10	15	1	1	(0.635,0.717)	(0.647,0.729)	(0.687,0.765)	(0.654,0.734)
10	15	1	3	(0.614,0.698)	(0.614,0.698)	(0.720,0.796)	(0.695,0.773)
10	15	1	5	(0.616,0.700	(0.629,0.711)	(0.784,0.852)	(0.735,0.809)
10	15	3	1	(0.716,0.792)	(0.660,0.740)	(0.683,0.761)	(0.643,0.725)
10	15	3	3	(0.699,0.777)	(0.693,0.771)	(0.737,0.811)	(0.654,0.734)
10	15	3	5	(0.737,0.811)	(0.687,0.765)	(0.786,0.854)	(0.697,0.775)
10	15	5	1	(0.759,0.829)	(0.729,0.803)	(0.652,0.732)	(0.649,0.731)
10	15	5	3	(0.791.0.857)	(0.725,0.799)	(0.763,0.833)	(0.706,0.782)
10	15	5	5	(0.778,0.846)	(0.731,0.805)	(0.773,0.843)	(0.706,0.782)
15	10	1	1	(0.656,0.736)	(0.649,0.731)	(0.670,0.750)	(0.629,0.711)
15	10	1	3	(0.668,0.748)	(0.662,0.742)	(0.708,0.784)	(0.664,0.744)
15	10	1	5	(0.645,0.727)	(0.656,0.736)	(0.759,0.829)	(0.660,0.740)
15	10	3	1	(0.718,0.794)	(0.710,0.786)	(0.608,0.692)	(0.592,0.676)
15	10	3	3	(0.742,0.814)	(0.706,0.782)	(0.704,0.780)	(0.685,0.763)
15	10	3	5	(0.704,0.780)	(0.689,0.767)	(0.782,0.850)	(0.708,0.784)
15	10	5	1	(0.797,0.863)	(0.733,0.807)	(0.614,0.698)	(0.596,0.680)
15	10	5	3	(0.795,0.861)	(0.691,0.769)	(0.727,0.801)	(0.666,0.746)
15	10	5	5	(0.791,0.857)	(0.752,0.824)	(0.759,0.829)	(0.710,0.786)
15	15	1	1	(0.637,0.719)	(0.643,0.725)	(0.633,0.715)	(0.604,0.688)
15	15	1	3	(0.625,0.707)	(0.641,0.723)	(0.739,0.813)	(0.681,0.759)
15	15	1	5	(0.676,0.756)	(0.689,0.767)	(0.797,0.863)	(0.704,0.780)
15	15	3	1	(0.761,0.831)	(0.689,0.767)	(0.662,0.742)	(0.649,0.731)
15	15	3	3	(0.750,0.822)	(0.706,0.782)	(0.776,0.844)	(0.691,0.769)
15	15	3	5	(0.750,0.822)	(0.702,0.778)	(0.821,0.833)	(0.746,0.818)
15	15	5	1	(0.795,0.861)	(0.759,0.829)	(0.674,0.754)	(0.656,0.736)
15	15	5	3	(0.793,0.859)	(0.761,0.831)	(0.756,0.828)	(0.693,0.771)
15	15	5	5	(0.784,0.852)	(0.739,0.813)	(0.784,0.852)	(0.693,0.771)

108

TABLE 3 (continued)

C. $\theta_1 = -0.75$ $\theta_2 = 0.75$

N1	N2	K1	K2	PK1	QK1	PK2	QK2
10	10	1	1	(0.662,0.742)	(0.654,0.734)	(0.623,0.705)	(0.602,0.686)
10	10	1	3	(0.645,0.727)	(0.643,0.725)	(0.687,0.765)	(0.670,0.750)
10	10	1	5	(0.629,0.711)	(0.643,0.725)	(0.739,0.813)	(0.708,0.784)
10	10	3	1	(0.725,0.799)	(0.676,0.756)	(0.668,0.748)	(0.643,0.725)
10	10	3	3	(0.695,0.773)	(0.662,0.742)	(0.739,0.813)	(0.654,0.734)
10	10	3	5	(0.737,0.811)	(0.689,0.767)	(0.756,0.828)	(0.660,0.740)
10	10	5	1	(0.784,0.852)	(0.710,0.786)	(0.627,0.709)	(0.588,0.672)
10	10	5	3	(0.776,0.844)	(0.702,0.778)	(0.725,0.799)	(0.643,0.725)
10	10	5	5	(0.744,0.816)	(0.716,0.792)	(0.718,0.794)	(0.674,0.754)
10	15	1	1	(0.602,0.686)	(0.594,0.678)	(0.641,0.723)	(0.641,0.723)
10	15	1	3	(0.654,0.734)	(0.660,0.740)	(0.729,0.803)	(0.679,0.757)
10	15	1	5	(0.616,0.700)	(0.616,0.700)	(0.825,0.887)	(0.748,0.820)
10	15	3	1	(0.754,0.826)	(0.704,0.780)	(0.662,0.742)	(0.635,0.717)
10	15	3	3	(0.716,0.792)	(0.687,0.765)	(0.725,0.799)	(0.695,0.773)
10	15	3	5	(0.737,0.811)	(0.666,0.746)	(0.786,0.854)	(0.695,0.773)
10	15	5	1	(0.756,0.828)	(0.676,0.756)	(0.662,0.742)	(0.647,0.729)
10	15	5	3	(0.742,0.814)	(0.687,0.765)	(0.742,0.814)	(0.683,0.761)
10	15	5	5	(0.759,0.829)	(0.714,0.790)	(0.817,0.879)	(0.731,0.805)
15	10	1	1	(0.674,0.754)	(0.664,0.744)	(0.637,0.719)	(0.637,0.719)
15	10	1	3	(0.672,0.752)	(0.695,0.773)	(0.727,0.801)	(0.639,0.721)
15	10	1	5	(0.633,0.715)	(0.635,0.717)	(0.746,0.818)	(0.718,0.794)
15	10	3	1	(0.754,0.826)	(0.691,0.769)	(0.676,0.756)	(0.660,0.740)
15	10	3	3	(0.729,0.803)	(0.683,0.761)	(0.683,0.761)	(0.647,0.729)
15	10	3	5	(0.761,0.831)	(0.689,0.767)	(0.778,0.846)	(0.718,0.794)
15	10	5	1	(0.759,0.829)	(0.742,0.814)	(0.649,0.731)	(0.641,0.723)
15	10	5	3	(0.806,0.870)	(0.750,0.822)	(0.710,0.786)	(0.660,0.740)
15	10	5	5	(0.795,0.861)	(0.710,0.786)	(0.780,0.848)	(0.714,0.790)
15	15	1	1	(0.645,0.727)	(0.645,0.727)	(0.629,0.711)	(0.629,0.711)
15	15	1	3	(0.658,0.738)	(0.681,0.759)	(0.695,0.773)	(0.631,0.713)
15	15	1	5	(0.681,0.759)	(0.687,0.765)	(0.810,0.874)	(0.733,0.807)
15	15	3	1	(0.771,0.841)	(0.725,0.799)	(0.604,0.688)	(0.567,0.653)
15	15	3	3	(0.748,0.820)	(0.727,0.801)	(0.727,0.801)	(0.664,0.744)
15	15	3	5	(0.723,0.797)	(0.668,0.748)	(0.756,0.828)	(0.706,0.782)
15	15	5	1	(0.786,0.854)	(0.720,0.796)	(0.641,0.723)	(0.614,0.698)
15	15	5	3	(0.793,0.859)	(0.725,0.799)	(0.697,0.775)	(0.652,0.732)
15	15	5	5	(0.769,0.839)	(0.737,0.811)	(0.797,0.863)	(0.718,0.794)

109

D. $\theta_1=-0.5$ $\theta_2=0.5$

N1	N2	K1	K2	PK1	QK1	PK2	QK2
10	10	1	1	(0.561,0.647)	(0.551,0.637)	(0.579,0.665)	(0.547,0.633)
10	10	1	3	(0.539,0.625)	(0.573,0.659)	(0.627,0.709)	(0.563,0.649)
10	10	1	5	(0.529,0.615)	(0.555,0.641)	(0.647,0.729)	(0.614,0.698)
10	10	3	1	(0.654,0.734)	(0.625,0.707)	(0.579,0.665)	(0.547,0.633)
10	10	3	3	(0.647,0.729)	(0.625,0.707)	(0.647,0.729)	(0.577,0.663)
10	10	3	5	(0.664,0.744)	(0.637,0.719)	(0.641,0.723)	(0.616,0.700)
10	10	5	1	(0.641,0.723)	(0.608,0.692)	(0.602,0.686)	(0.553,0.639)
10	10	5	3	(0.641,0.723)	(0.645,0.727)	(0.614,0.698)	(0.584,0.668)
10	10	5	5	(0.660,0.740)	(0.631,0.713)	(0.652,0.732)	(0.639,0.721)
10	15	1	1	(0.582,0.666)	(0.545,0.631)	(0.533,0.619)	·(0,0.553,0.639)
10	15	1	3	(0.567,0.653)	(0.590,0.674)	(0.637,0.719)	(0.590,0.674)
10	15	1	5	(0.582,0.666)	(0.579,0.665)	(0.704,0.780)	(0.685,0.763)
10	15	3	1	(0.610,0.694)	(0.582,0.666)	(0.588,0.672)	(0.551,0.637)
10	15	3	3	(0.629,0.711)	(0.590,0.674)	(0.668,0.748)	(0.606,0.690)
10	15	3	5	(0.602,0.686)	(0.575,0.661)	(0.672,0.752)	(0.606,0.690)
10	15	5	1	(0.647,0.729)	(0.627,0.709)	(0.573,0.659)	(0.559,0.645)
10	15	5	3	(0.639,0.721)	(0.631,0.713)	(0.656,0.736)	(0.629,0.711)
10	15	5	5	(0.639,0.721)	(0.621,0.703)	(0.689,0.767)	(0.616,0.700)
15	10	1	1	(0.616,0.700)	(0.594,0.678)	(0.559,0.645)	(0.543,0.629)
15	10	1	3	(0.582,0.666)	(0.584,0.668)	(0.600,0.684)	(0.575,0.661)
15	10	1	5	(0.590,0.674)	(0.575,0.661)	(0.621,0.703)	(0.600,0.684)
15	10	3	1	(0.629,0.711)	(0.590,0.674)	(0.549,0.635)	(0.535,0.621)
15	10	3	3	(0.656,0.736)	(0.645,0.727)	(0.616,0.700)	(0.579,0.665)
15	10	3	5	(0.631,0.713)	(0.606,0.690)	(0.670,0.750)	(0.641,0.723)
15	10	5	1	(0.710,0.786)	(0.645,0.727)	(0.541,0.627)	(0.529,0.615)
15	10	5	3	(0.668,0.748)	(0.641,0.723)	(0.610,0.694)	(0.553,0.639)
15	10	5	5	(0.710,0.786)	(0.656,0.736)	(0.652,0.732)	(0.604,0.688)
15	15	1	1	(0.571,0.657)	(0.571,0.657)	(0.590,0.674)	(0.565,0.651)
15	15	1	3	(0.521,0.607)	(0.549,0.635)	(0.633,0.715)	(0.569,0.655)
15	15	1	5	(0.610,0.694)	(0.608,0.692)	(0.662,0.742)	(0.582,0.666)
15	15	3	1	(0.660,0.740)	(0.629,0.711)	(0.604,0.688)	(0.573,0.659)
15	15	3	3	(0,625,0.707)	(0.614,0.698)	(0.687,0.765)	(0.602,0.686)
15	15	3	5	(0.631,0.713)	(0.598,0.682)	(0.689,0.767)	(0.614,0.698)
15	15	5	1	(0.674,0.754)	(0.612,0.696)	(0.596,0.680)	(0.575,0.661)
15	15	5	3	(0.693,0.771)	(0.641,0.723)	(0.691,0.769)	(0.584,0.668)
15	15	5	5	(0.695,0.773)	(0.639,0.721)	(0.654,0.734)	(0.610,0.694)

E. $\theta_1 = -0.3$ $\theta_2 = 0.2$

N1	N2	K1	K2	PK1	QK1	PK2	QK2
10	10	1	1	(0.456,0.544)	(0.458,0.546)	(0.504,0.592)	(0.488,0.576)
10	10	1	3	(0.482,0.570)	(0.498,0.586)	(0.470,0.558)	(0.458,0.546)
10	10	1	5	(0.478,0.566)	(0.482,0.570)	(0.490,0.578)	(0.502,0.590)
10	10	3	1	(0.506,0.594)	(0.525,0.611)	(0.480,0.568)	(0.470,0.558)
10	10	3	3	(0.496,0.584)	(0.494,0.582)	(0.533,0.619)	(0.446,0.534)
10	10	3	5	(0.523,0.609)	(0.490,0.578)	(0.537,0.623)	(0.557,0.643)
10	10	5	1	(0.490,0.578)	(0.508,0.596)	(0.527,0.613)	(0.508,0.596)
10	10	5	3	(0.502,0.590)	(0.506,0.594)	(0.486,0.574)	(0.484,0.572)
10	10	5	5	(0.470,0.558)	(0.468,0.556)	(0.502,0.590)	(0.521,0.607)
10	15	1	1	(0.460,0.548)	(0.472,0.560)	(0.523,0.609)	(0.504,0.592)
10	15	1	3	(0.454,0.542)	(0.486,0.574)	(0.565,0.651)	(0.539,0.625)
10	15	1	5	(0.506,0.594)	(0.488,0.576)	(0.579,0.665)	(0.533,0.619)
10	15	3	1	(0.474,0.562)	(0.488,0.576)	(0.508,0.596)	(0.498,0.586)
10	15	3	3	(0.521,0.607)	(0.514,0.602)	(0.523,0.609)	(0.525,0.611)
10	15	3	5	(0.494,0.582)	(0.462,0.550)	(0.559,0.645)	(0.521,0.607)
10	15	5	1	(0.482,0.570)	(0.468,0.556)	(0.482,0.570)	(0.478,0.566)
10	15	5	3	(0.514,0.602)	(0.521,0.607)	(0.525,0.611)	(0.490,0.578)
10	15	5	5	(0.506,0.594)	(0.521,0.607)	(0.567,0.653)	(0.510,0.598)
15	10	1	1	(0.504,0.592)	(0.512,0.600)	(0.432,0.520)	(0.458,0.546)
15	10	1	3	(0.551,0.637)	(0.539,0.625)	(0.500.0.588)	(0.470,0.558)
15	10	1	5	(0.531,0.617)	(0.533,0.619)	(0.512,0.600)	(0.527,0.613)
15	10	3	1	(0.504,0.592)	(0.476,0.564)	(0.470,0.558)	(0.452,0.540)
15	10	3	3	(0.510,0.598)	(0.498,0.586)	(0.470,0.558)	(0.448,0.536)
15	10	3	5	(0.545,0.631)	(0.519,0.605)	(0.500,0.588)	(0.510,0.598)
15	10	5	1	(0.561,0.647)	(0.521,0.607)	(0.504,0.592)	(0.512,0.600)
15	10	5	3	(0.592,0.676)	(0.573,0.659)	(0.480,0.568)	(0.450,0.538)
15	10	5	5	(0.567,0.653)	(0.547,0.633)	(0.512,0.600)	(0.464,0.552)
15	15	1	1	(0.482,0.570)	(0.486,0.574)	(0.508,0.596)	(0.504,0.592)
15	15	1	3	(0.468,0.556)	(0.474,0.562)	(0.529,0.615)	(0.496,0.584)
15	15	1	5	(0.464,0.552)	(0.472,0.560)	(0.547,0.633)	(0.498,0.586)
15	15	3	1	(0.516,0.604)	(0.514,0.602)	(0.502,0.590)	(0.504,0.592)
15	15	3	3	(0.533,0.619)	(0.508,0.596)	(0.543,0.629)	(0.525,0.611)
15	15	3	5	(0.541,0.627)	(0.547,0.633)	(0.537,0.623)	(0.523,0.609)
15	15	5	1	(0.567,0.653)	(0.519,0.605)	(0.472,0.560)	(0.446,0.534)
15	15	5	3	(0.533,0.619)	(0.531,0.617)	(0.559,0.645)	(0.551,0.637)
15	15	5	5	(0.535,0.621)	(0.539,0.625)	(0.551,0.637)	(0.504,0.592)

ACKNOWLEDGMENT

We are very grateful to Howard Carson who has taken enormous trouble both in programming and actually computing the details which are summarized in Tables 1, 2 and 3.

The author Kamal C. Chanda is now at Texas Tech University, Lubbock, Texas.

APPENDIX

A.1. PROOF OF THEOREM 1

We can write

$$V = n_1^{-1} \sum_{i=1}^{n_1} H(X_i) + n_2^{-1} \sum_{i=1}^{n_2} H(Y_i) \qquad (A.1.1)$$

where

$$H(x) = \sum_{j=1}^{k} \phi(x, Z_j)/k, \quad \phi(x,y) = \phi(x-y) \qquad (A.1.2)$$

$$\phi(u) = \begin{cases} 0 & \text{if } u>0 \\[2mm] \frac{1}{2} & \text{if } u<0 \end{cases} \qquad (A.1.3)$$

Define

$$U = W-\nu, \quad \nu = E(W) = E[\phi(X_1-Y_1)] = \int F_2 dF_1 \qquad (A.1.4)$$

$$U^* = n_1^{-1} \sum_{i=1}^{n_1} [\psi_{10}(X_i)-\nu] + n_2^{-1} \sum_{i=1}^{n_2} [\psi_{01}(Y_i)-\nu] \qquad (A.1.5)$$

where

$$\psi_{10}(x) = E[\psi(x,Y_1)] \qquad (A.1.6)$$

$$\psi_{01}(y) = E[\psi(X_1,y)] \qquad (A.1.7)$$

$$\psi(x,y) = \psi(x-y), \quad \psi(u) = 1-2\phi(u) \qquad (A.1.8)$$

Let

$$m = \min(n_1, n_2) \tag{A.1.9}$$

$$n = n_1 n_2 / (n_1 + n_2) \tag{A.1.10}$$

Since U and U^* are both bounded

$$E[U^*-U)^{2r}] = o(m^{-2r}) \tag{A.1.11}$$

for any $r = 0, 1, 2, .$ [See Grams and Serfling (1973)]. Let

$$\pi = \frac{1}{2} - \nu \tag{A.1.12}$$

We assume that $\pi \neq 0$.

Note that $W > \dfrac{1}{2}$ <=> $U > \pi$.

<u>Case 1.</u> $\pi > 0$ <=> $\nu < \dfrac{1}{2}$ (if $F_1 > F_2$, $\nu < \dfrac{1}{2}$). Observe that

$$P(W>\tfrac{1}{2},\ V>\tfrac{1}{2}) < P(U>\pi) \tag{A.1.13}$$

$$\begin{aligned}
P(U>\pi) &\leq P(|U|>\pi) \\
&= P(|U-U^*+U^*|>\pi) \\
&\leq P(|U-U^*|>\pi/2) + P(|U^*>\pi/2)
\end{aligned} \tag{A.1.14}$$

$$P(|U-U^*|>\pi/2) \leq 4^r \pi^{-2r} E[(U-U^*)^{2r}] = o(m^{-2r}) \tag{A.1.15}$$

for any $r = 0, 1, 2, \ldots$

$$P(|U^*|>\pi/2) = P\left(\left|n_1^{-1} \sum_{i=1}^{n_1} X_i^* + n_2^{-1} \sum_{i=1}^{n_2} Y_i^*\right| > \pi/2\right)$$

$$\leq P\left(\left|n_1^{-1} \sum_{i=1}^{n_1} X_i^*\right| > \pi/4\right) + P\left(\left|n_2^{-1} \sum_{i=1}^{n_2} Y_i^*\right| > \pi/4\right) \tag{A.1.16}$$

where

$$X_i^* = \psi_{10}(X_i) - \nu, \quad Y_i^* = \psi_{01}(Y_i) - \nu \tag{A.1.17}$$

(Note that $E[\psi_{10}(X_1)] = E[\psi_{01}(Y_1)] = \nu$.) Since X_1^*, Y_1^* are bounded r.v.'s which can take both positive and negative values, the moment generating function of X_1^*, Y_1^* exists, and therefore, we can find ρ_1, ρ_2 $(0 < \rho_i < 1$, $i = 1, 2)$ (see Hoeffding (1963)), such that

$$P\left(\left|n_1^{-1} \Sigma X_i^*\right| > \pi/4\right) = O(\rho_1^{n_1}) \tag{A.1.18}$$

$$P\left(\left|n_2^{-1} \Sigma Y_i^*\right| > \pi/4\right) = O(\rho_2^{n_2}) \tag{A.1.19}$$

Combining (A.1.14)-(A.1.19) we have, therefore

$$P(W > \tfrac{1}{2}) = P(U > \pi) = o(m^{-r}) \tag{A.1.20}$$

for any $r = 0, 1, 2, \ldots$ From (A.1.13) and (A.1.20) we have

$$P(W \leq \tfrac{1}{2}, \quad V \leq \tfrac{1}{2}) = P(V \leq \tfrac{1}{2}) - P(W > \tfrac{1}{2}, \quad V \leq \tfrac{1}{2})$$

$$= P(V \leq \tfrac{1}{2}) + o(m^{-r}) \tag{A.1.21}$$

for any $r = 0, 1, 2, \ldots$ Therefore, to any order m^{-r}

$$Q_k = P(V \leq \tfrac{1}{2}) \tag{A.1.22}$$

Again

$$P(V \leq \tfrac{1}{2} \mid \underset{\sim}{Z} = \underset{\sim}{z}) = P\left(n_1^{-1} \sum_{i=1}^{n_1} X_i^{**} + n_2^{-1} \sum_{i=1}^{n_2} Y_i^{**} \leq p(\underset{\sim}{z})\right) \tag{A.1.23}$$

where

$$X_i^{**} = k^{-1} \sum_1^k \phi(X_i, z_j) - \nu_1(\underset{\sim}{z}) \tag{A.1.24}$$

$$Y_i^{**} = k^{-1} \sum_1^k \phi(Y_i, z_j) - \nu_2(\underset{\sim}{z}) \tag{A.1.25}$$

$$\nu_1(\underset{\sim}{z}) = (2k)^{-1} \sum_{j=1}^k F_1(z_j), \tag{A.1.26}$$

$$\nu_2(z) = (2k)^{-1} \sum_{j=1}^{k} F_2(z_j) \tag{A.1.27}$$

$$p(\underset{\sim}{z}) = \frac{1}{2} - \nu_1(\underset{\sim}{z}) - \nu_2(\underset{\sim}{z})$$

$$= \sum_{j=1}^{k} [1-F_1(z_j) - F_2(z_j)]/2k \tag{A.1.28}$$

Note that $\left|X_i^{**}\right| \leq k/2$, $\left|Y_i^{**}\right| \leq k/2$ for all i. For any fixed $\varepsilon > 0$ consider the set of $\underset{\sim}{z}$ for which

$$p(\underset{\sim}{z}) < -\varepsilon \tag{A.1.29}$$

Then replacing X_i^{**} by $-X_i^{**}$ and Y_i^{*} by $-Y_i^{**}$ whereby the inequality on the r.h.s. of (A.1.23) is reversed with $p(z)$ replaced by $-p(\underset{\sim}{z}) > \varepsilon$, and making an appeal to Theorem 2 in Hoeffding (1963), we have

$$P(V \leq \tfrac{1}{2} \mid \underset{\sim}{Z} = \underset{\sim}{z}) \leq e^{-n\varepsilon^2/k^2} \tag{A.1.30}$$

for any $\underset{\sim}{z}$ for which $p(\underset{\sim}{z}) < -\varepsilon$, where $n = n_1 n_2/(n_1 + n_2)$. Consider now a $\underset{\sim}{z}$ for which $p(\underset{\sim}{z}) > \varepsilon$. Then

$$1 - P(V \leq \tfrac{1}{2} \mid \underset{\sim}{Z} = \underset{\sim}{z}) = P\left(n_1^{-1} \sum_{1}^{n_1} X_i^{**} + n_2^{-1} \sum_{1}^{n_2} Y_i^{**} > p(z)\right)$$

$$\leq P\left(n_1^{-1} \sum_{1}^{n_1} X_i^{**} + n_2^{-1} \sum_{1}^{n_2} Y_i^{**} > \varepsilon\right)$$

$$\leq e^{-n\varepsilon^2/k^2} \tag{A.1.31}$$

again by Theorem 2 in Hoeffding (1963). Since

$$\chi(z) = F_1(z) + F_2(z) - 1 \tag{A.1.32}$$

is a continuous monotone function of z

$$P[-\varepsilon \leq p(\underset{\sim}{Z}) \leq \varepsilon] = P[-2k\varepsilon \leq \sum_j \chi(Z_j) \leq 2k\varepsilon] < N\varepsilon \tag{A.1.33}$$

where we use N as a generic symbol to denote a positive constant which depends on k only. Combining (A.1.30), (A.1.31), (A.1.33), the fact that

$$P(V \leq \tfrac{1}{2}) = E[P(V \leq \tfrac{1}{2}|\underset{\sim}{Z})] \tag{A.1.34}$$

choosing $\varepsilon = o(n^{-\alpha})$ where $\alpha < \tfrac{1}{2}$ and after some simplification we conclude that

$$\left| P(V \leq \tfrac{1}{2}) - Q_k^0 \right| \leq Nn^{-\alpha} \tag{A.1.35}$$

where

$$Q_k^0 = P\left[\sum_{j=1}^{k} \chi(Z_j) < 0 \right] \tag{A.1.36}$$

Now use (A.1.22) and (A.1.35) to derive the result of the theorem.

A.2.   PROOF OF THEOREM 2.

It is easy to see that if $\Delta > 0$

$$P_k = P(S_1 S_2 > 0) \tag{A.2.1}$$

where

$$S_1 = \hat{\theta}_1 - \hat{\theta}_2 = \theta_1^* - \theta_2^* - 2\Delta \tag{A.2.2}$$

$$S_2 = R - (\theta_1^* + \theta_2^*)/2 \tag{A.2.3}$$

$$\theta_j^* = \hat{\theta}_j - \theta_j \tag{A.2.4}$$

$$\Delta = (\theta_2 - \theta_1)/2 \tag{A.2.5}$$

$$R = k^{-1} \sum_{j=1}^{k} [Z_j - (\theta_1 + \theta_2)/2] \tag{A.2.6}$$

with $\hat{\theta}_j$ being the consistent estimator of θ_j. We first assume that $\hat{\theta}_1 = \overline{X}$, $\hat{\theta}_2 = \overline{Y}$ so that

$$S_1 \sim N(-2\Delta, \sigma_1^2)$$

$$\tag{A.2.7}$$

$$S_2 \sim N(-\delta, \sigma_2^2)$$

where

$$\delta = (\theta_1 + \theta_2)/2 - \theta \qquad\qquad (A.2.8)$$

$$\sigma_1^2 = n_1^{-1} + n_2^{-1} = n^{-1} \qquad\qquad (A.2.9)$$

$$\sigma_2^2 = k^{-1} + (4n)^{-1} \qquad\qquad (A.2.10)$$

Note that $S_1 S_2 > 0$ is equivalent to either $S_1 > 0$, $S_2 > 0$ or $S_1 < 0$, $S_2 < 0$. Hence if we standardize S_j by S_j^* where $S_1^* = (S_1 + 2\Delta)/\sigma_1$, $S_2^* = (S_2 + \delta)/\sigma_1$, then

$$P_k = P(S_1^* > 2\Delta/\sigma_1, \ S_2^* > \delta/\sigma_2)$$

$$+ \ P(S_1^* < 2\Delta/\sigma_1, \ S_2^* < \delta/\sigma_2) \qquad (A.2.11)$$

Again the first expression on the r.h.s. of (A.2.11) is

$$\leq P(S_1^* > 2\Delta/\sigma_1) = 1 - \Phi(2\Delta/\sigma_1) = O(\rho^n) \qquad (A.2.12)$$

for some ρ $(0 < \rho < 1)$. Similarly the absolute difference between the second expression in (A.2.11) and $P(S_2^* < \delta/\sigma_2)$ is

$$\leq P(S_1^* > 2\Delta/\sigma_1) = O(\rho^n) \qquad\qquad (A.2.13)$$

Also

$$P(S_2^* < \delta/\sigma_2) = \Phi(\delta/\sigma_2)$$

$$= \Phi(\delta\sqrt{k}) - \delta k^{\frac{3}{2}} \phi(\delta\sqrt{k})/8n + o(n^{-1}). \qquad (A.2.14)$$

Combining (A.2.11)-(A.2.14) we have, therefore, the result of the theorem when $\Delta > 0$. Similarly we can prove the other result for $\Delta < 0$. The proof of (4.34) is easy and hence omitted.

BIBLIOGRAPHY

Anderson, T. W.   (1958).   *An Introduction to Multi-variate Statistical Analysis*.   New York: John Wiley and Sons, Inc.

Barnett, V. D.   (1966).   Order statistics estimators of
     the location of the Cauchy distribution.   *J. Amer.
     Statist. Assoc.* 61, 1205-18.

Chernoff, H., Gastwirth, J. L. and Johns, M. V., Jr.   (1967).
     Asymptotic distribution of linear combinations of
     functions of order statistics with applications to
     estimation.   *Ann. Math. Statist.* 38, 52-72.

Cochran, W. G. and Hopkins, C. D.   (1961).   Some
     classification problems with multivariate qualitive
     data.   *Biometrics* 17, 10-32.

Das Gupta, Somesh   (1964).   Nonparametric classification
     rules.   *Sankhyā,* Ser. A 26, 25-30.

Fix, E., and Hodges, J. L.   (1952).   Discriminatory
     analysis: nonparametric discrimination, small
     sample performances.   Report No. 11, School of
     Aviation Medicine, Randolph Air Force Base, Texas.

Fix, E., and Hodges, J. L.   (1959).   Discriminatory
     analysis: nonparametric discrimination, consistency
     properties.   Report No. 4, School of Aviation
     Medicine, Randolph Air Force Base, Texas.

Govindarajulu, Z. and Gupta, A. K.   (1972).   Certain
     non-parametric classification rules: univariate
     case.   Tech. Report 7, Department of Statistics,
     University of Michigan.

Grams, W. F. and Serfling, R. J.   (1973).   Convergence
     rates for U-statistics and related statistics.
     Ann. Statist. 1, 153-60.

Gupta, S. S. and Waknis, M. N.   (1964).   Estimation of
     parameters of the Logistic distribution.   Tech.
     Report 15, Department of Statistics, Purdue
     University.

Hoeffding, W.   (1963).   Probability inequalities for sums
     of bounded random variables.   *J. Amer. Statist.
     Assoc.* 58, 13-30.

Hudimoto, H. (1964). On a distribution-free two-way
classification. *Ann. Inst. Statist. Math.* 16,
247-53.

Jung, J. (1955). On linear estimates defined by a
continuous weight function. *Arkiv. fur Matematik*
3, 199-209.

Lachenbruch, P. A., Sneeringer, C. and Revo, L. T.
(1973). Robustness of the linear and quadratic
discriminant function to certain types of non-
normality. *Comm. Statist.* 1, 39-56.

Rao, C. R. (1965). *Linear Statistical Inference and Its
Applications.* New York: John Wiley and Sons, Inc.

Sorum, M. (1973). Estimating the expected probability
of misclassification for a rule based on linear
discriminant function: univariate normal case.
Technometrics 15, 329-40.

Stoller, D. C. (1954). Univariate two-population distribution-
free discrimination. *J. Amer. Statist. Assoc.* 49, 770-5.

SOME THOUGHTS ON PATTERN CLASSIFICATION BY ATTRIBUTES

G. P. Steck

Sandia Laboratories
Albuquerque, New Mexico

1. THE PATTERNS

Once it was decided to use handwritten chinese characters
to check the feasibility of the idea of letting features be defined by n-tuples with high transmission rates, I asked Dr.
Peter Chen (Sandia Laboratories) if he could spend a few
minutes a day marking 20 copies of each of 20 characters with
a felt pen on 3×5 index cards. The only instructions given
were to use a variety of character types including some degree
of "look-alike". All replicates of a character were made at
one time in sequence. All are legitimate characters except
one (I found out much later) which was created as a "look-
alike".

The complete data set is reproduced in Figures 1a - 1e.
Note the variation in line thickness and contrast. One repli-
cate in the bottom row of Figure 1b is retouched to eliminate
the feathery character of some of its lines which prevents the
proper operation of the scanner.

These patterns have been digitized by the SID-2 (Sandia
Image Digitizer, Mod 2) by recording the locations of black/
white and white/black transitions on a 500 × 500 grid. These
digitizations are "filled in" and preprocessed by centering and
scaling until the pattern touches at least one vertical edge
and at least one horizontal edge.

121

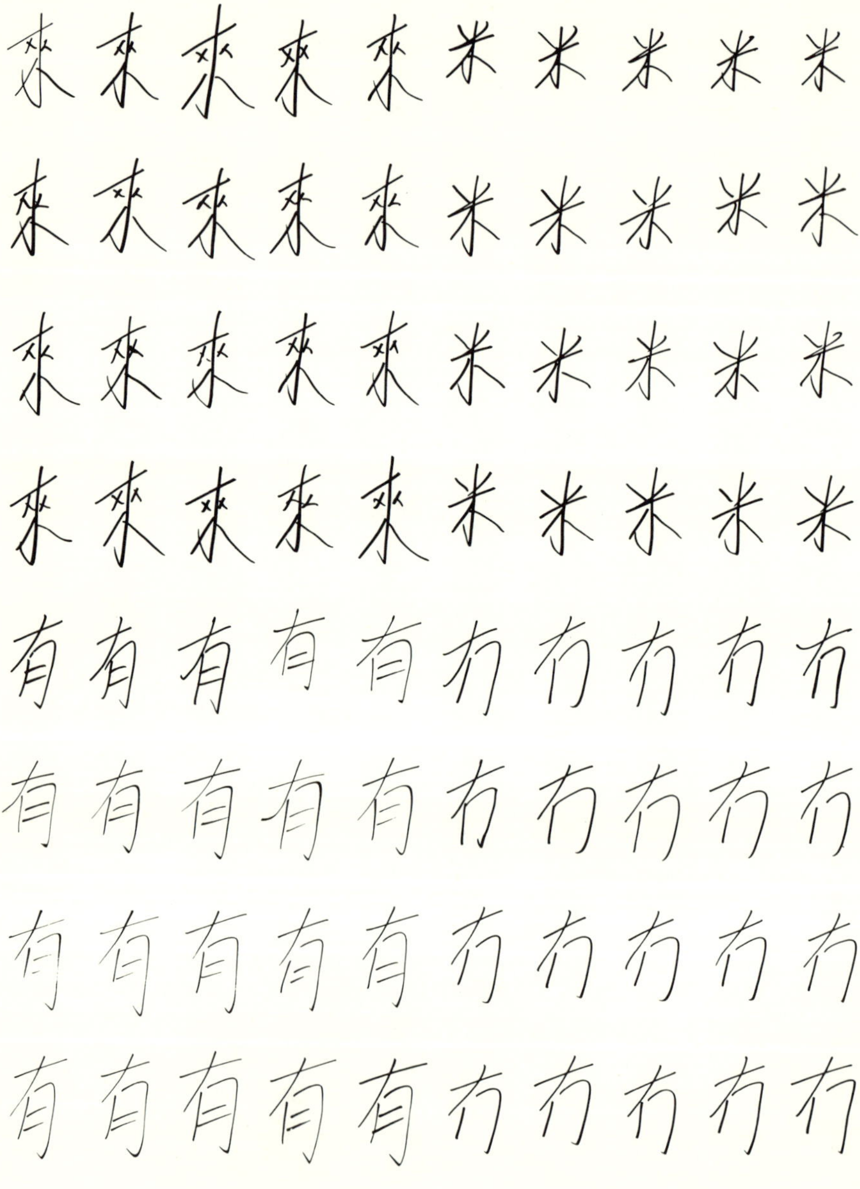

Figure 1a. Original characters: AB
                              CD

Figure 1b. Original characters: EF
GH

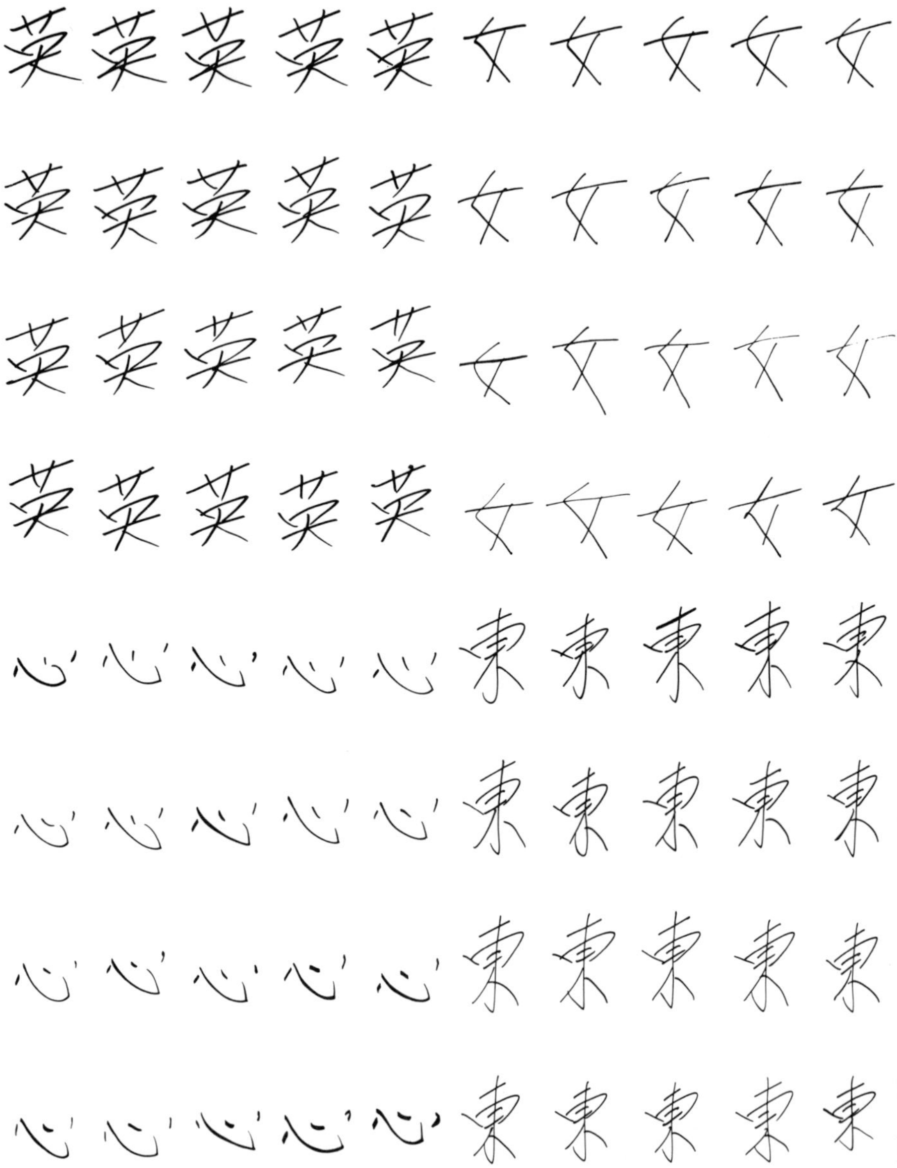

Figure 1c. Original characters: IJ
                                   KL

Figure 1d. Original characters: MN
                                  OP

Figure 1e. Original characters: QR
                                 ST

Since a 500 × 500 array is impossibly large and, fortunate-
ly, also unnecessary, the patterns are reduced to a more
manageable 10 × 10 array. This is done in two ways. One is
simple-minded and quick, the other is much cleverer but takes
much longer; however, both give about the same results.

Consider the 500 × 500 array to consist of 0's and
1's . To reduce it uniformly to a 10 × 10 array means 100
50 × 50 subarrays must each be converted to a 0 or a 1.
The simple-minded conversion is to say 1 if and only if more
than some suitable number of the 2500 possible 1's are
present. [After some preliminary searching I settled on
2500/30 as the suitable number.] The fraction is, of course,
the same for each point. The sophisticated conversion is to
choose the cutoff number separately for each point as that value
which maximizes the information transmitted by that point considered as a channel. Considering the change in the entropy
of a point, as the cutoff number increases to change the status
of a pattern, enables an algorithm to be derived which minimizes the entropy without computing all 399 of them.

Another approach to reduce the size of the mosaic, not
considered here, is to divide it into zones of more or less
equal information transmission rate instead of equal area.

The result of reducing the patterns to a 10 × 10 mosaic
by the sophisticated method is shown in Figures 2a to 2e.
Note how much detail is lost. Figure 3 shows the degree of
fuzziness of the pattern classes.

2.   FEATURE SELECTION

Now that the original patterns are reduced to a square
array of 100 0's and 1's we can exploit the n-tuple method
of Browning and Bledsoe (1959). They select as features
ordered sets of n randomly chosen mosaic elements (n-tuple).

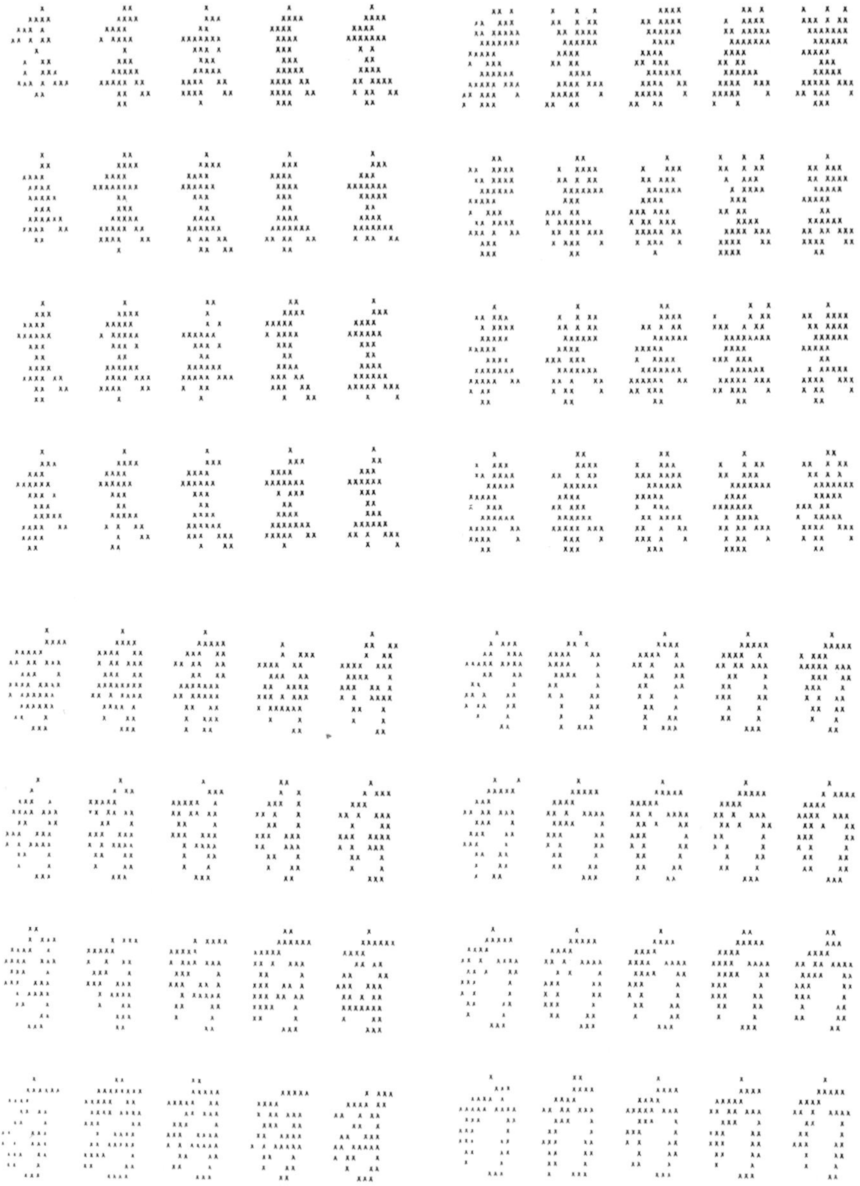

Figure 2a. Digitized characters: AB
                                   CD

Figure 2b. Digitized characters: EF
GH

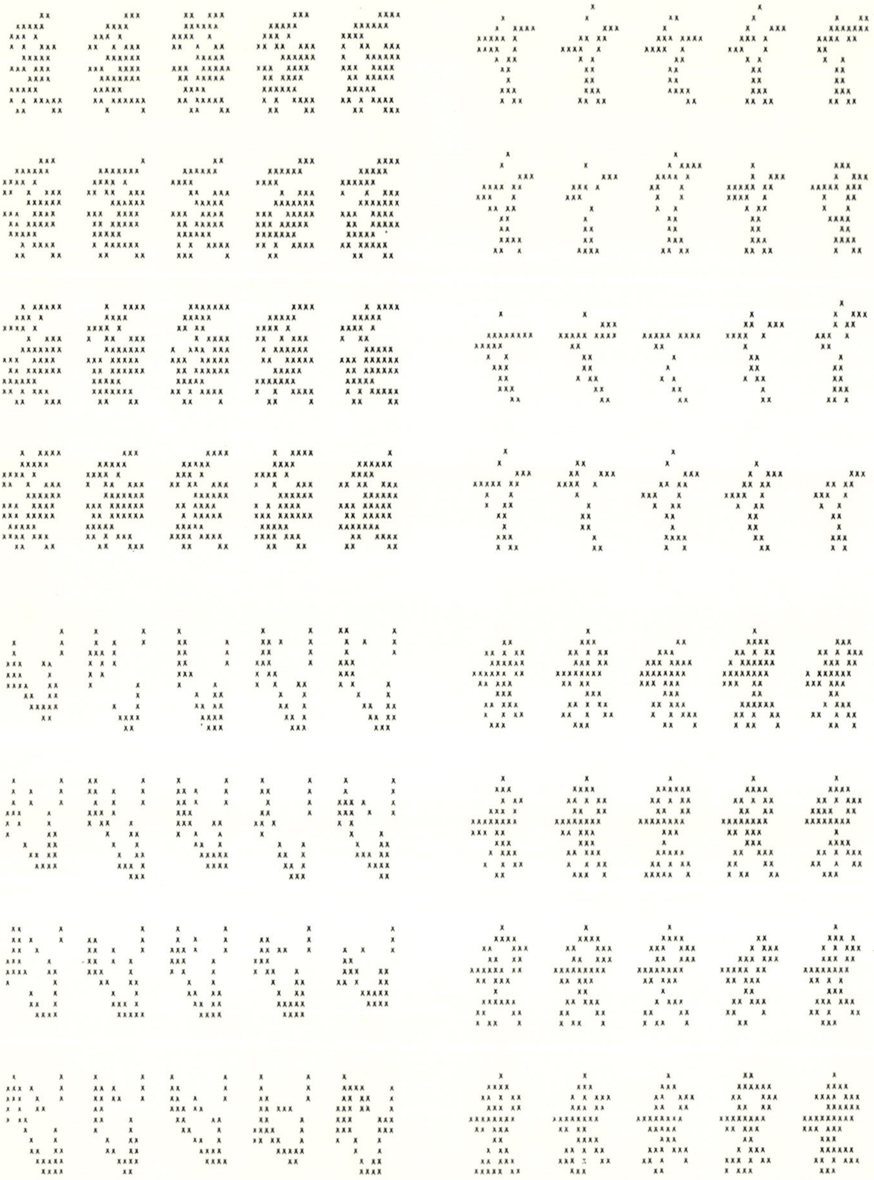

Figure 2c. Digitized characters: IJ
KL

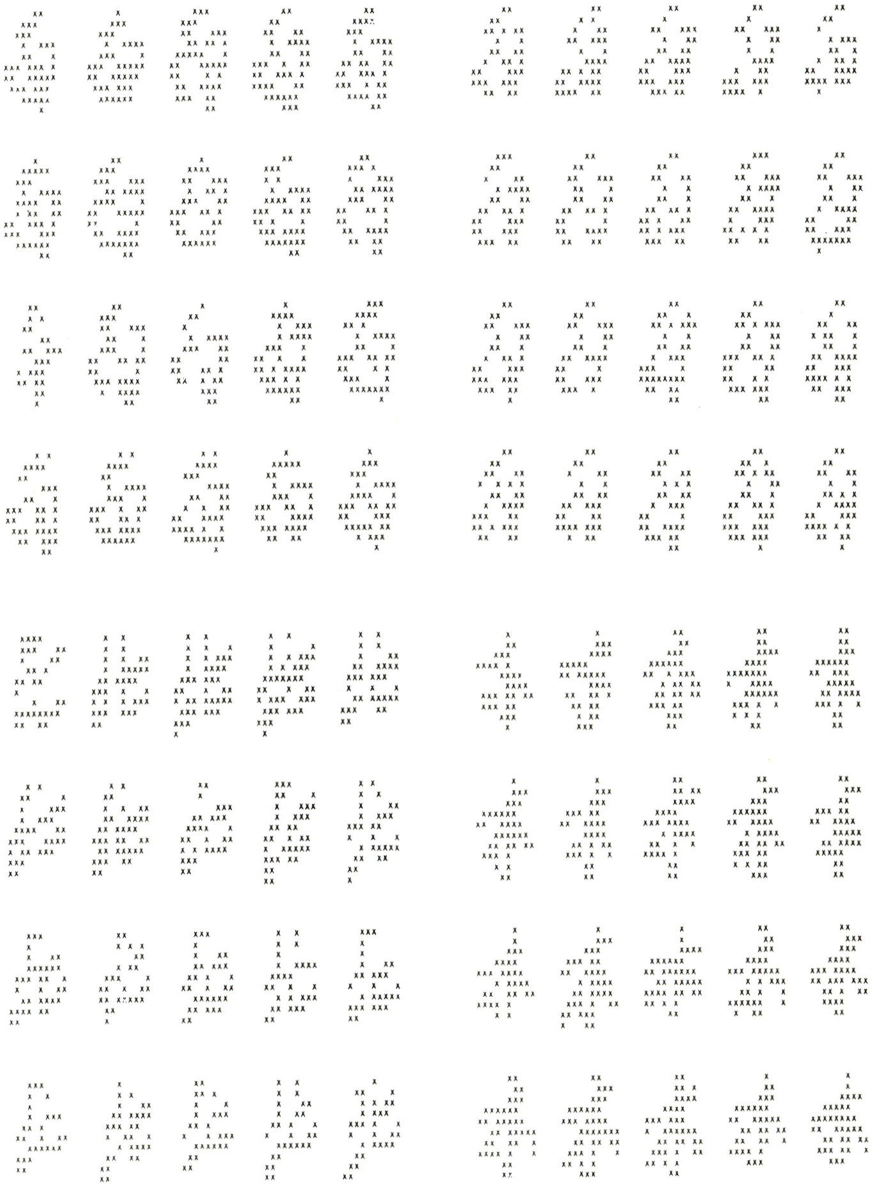

Figure 2d. Digitized characters: MN
OP

Figure 2e. Digitized characters: QR
                                 ST

blank - 0 to 5 occurrences
  0   - 6 to 10 occurrences
  ▨   - 11 to 15 occurrences
  ▩   - 16 to 20 occurrences

Figure 3. Fuzziness of pattern classes.

The appearance or state of this feature is the binary number
given by the n-tuple. The same thing is done here except that
we choose the points of an n-tuple from consecutive entries in
the ordering of mosaic elements by their information transmission rates.

Each point of the mosaic can be thought of as an information channel and as such has an average transmission rate (ATR).
For any point we have a sending alphabet of twenty chinese
characters and a receiving alphabet of two characters, 0
and 1. In general, however, for a sending alphabet, X_1, X_2,
..., X_s , and a receiving alphabet, Y_1, Y_2, ..., Y_r, we have
the channel matrix given in Table 1 where p_{ij} is the un-
conditional probability of sending X_i and receiving Y_j.

TABLE 1

Channel Matrix

	Send X		
	X_1 X_2 ... X_s	row sums
Y_1	p_{11} p_{21} ... p_{s1}	π_1
Y_2	p_{12} p_{22} ... p_{s2}	π_2
$\vdots$	$\vdots$ $\vdots$ $\vdots$	$\vdots$
Y_r	p_{1r} p_{2r} ... p_{sr}	π_r

(Receive Y)

The transmission rate (TR) for an individual row or state
of the channel matrix is [See, e.g., Feinstein (1958, Ch. 3),
for details.]

$$TR = \log s + \sum_{i=1}^{s} (p_{ij}/\pi_j) \log (p_{ij}/\pi_j) \qquad (2.1)$$

where all logarithms are to the base 2. Averaging this over
rows (multiply by π_j and sum) gives the Average Transmission Rate

$$\text{ATR} = \log s - \sum_{j=1}^{r} \pi_j \log \pi_j + \sum_{i=1}^{s} \sum_{j=1}^{r} p_{ij} \log p_{ij} \qquad (2.2)$$

In this application the p_{ij} and π_j are not known so we
estimate them by n_{ij}/N and m_j/N, respectively, where

$\quad n_{i1}$ = number of times array element is 0 for a character
$\qquad\quad$ of class i

$\quad n_{i2}$ = number of times array element is 1 for a character
$\qquad\quad$ of class i

$$m_j = \sum_{i=1}^{r} n_{ij} \qquad (2.3)$$

N = total number of characters in learning set = $\sum_{i} \sum_{j} n_{ij}$

This gives (approximately)

$$\text{ATR} \sim \log s - \sum_{j=1}^{r} m_j \log m_j + \sum_{i=1}^{s} \sum_{j=1}^{r} n_{ij} \log n_{ij} \qquad (2.4)$$

where the convention is $0 \log 0 \equiv 1$.

When the ATR is computed for each point of the 10 × 10
array we find, as expected, that boundary points are not, in
general, as informative as interior points.

Computing the ATR for each point of the array enables
ordering of the points according to how informative they are.
Now we can take the first n points in this ordering as the
first n-tuple; the next n points as the second n-tuple;
and so forth. (While this certainly orders n-tuples by in-
formativeness when n=1, it does not for $n \geq 2$, unfortunately; however, no attempt is made in this paper to search
for the most informative n-tuples, $n \geq 2$.)

These n-tuples are the features. This method of feature selection is adaptive in the sense that it depends on the alphabet and on the learning set.

When the channel matrix of Table 1 has p_{ij} replaced by n_{ij} we call it the state matrix. Tables 2abc give state matrices for some points when the learning set consists of all patterns. The notation (r,c) refers to the point in the rth row and c^{th} column of the pattern array.

TABLE 2a

State Matrices for (5,1)

Character Class	State 0	State 1
A	20	0
B	20	0
C	19	1
D	20	0
E	11	9
F	16	4
G	0	20
H	11	9
I	18	2
J	11	9
K	6	14
L	0	20
M	20	0
N	20	0
O	20	0
P	1	19
Q	4	16
R	2	18
S	0	20
T	0	20
TR	.5566	.7691

500×500 to 10×10 sophisticated reduction. ATR = .6528 bits. (most informative point)

Character Class	State 0	State 1
A	20	0
B	20	0
C	19	1
D	20	0
E	12	8
F	16	4
G	0	20
H	12	8
I	18	2
J	11	9
K	6	14
L	0	20
M	20	0
N	20	0
O	20	0
P	3	17
Q	4	16
R	2	18
S	0	20
T	0	20
TR	.5244	.7760

500×500 to 10×10 simple-minded reduction. ATR = .6357 bits. (most informative point)

TABLE 2b

State Matrices for (5,6)

		State	
		0	1
Character Class	A	1	19
	B	0	20
	C	20	0
	D	20	0
	E	19	1
	F	12	8
	G	18	2
	H	6	14
	I	0	20
	J	20	0
	K	20	0
	L	1	19
	M	14	6
	N	19	1
	O	3	17
	P	1	19
	Q	14	6
	R	0	20
	S	18	2
	T	18	2
TR		.5424	.7560

		State	
		0	1
Character Class	A	0	20
	B	0	20
	C	19	1
	D	14	6
	E	2	18
	F	9	11
	G	2	18
	H	1	19
	I	0	20
	J	7	13
	K	17	3
	L	0	20
	M	0	20
	N	11	9
	O	0	20
	P	0	20
	Q	5	15
	R	0	20
	S	12	8
	T	14	6
TR		1.0563	.1828

500×500 to 10×10 sophisticated
reduction. ATR = .6364 bits
(second most informative
point)

500×500 to 10×10 simple-minded
reduction. ATR = .4296 bits
(forty-seventh most informative
point)

TABLE 2c

State Matrix for $\{(5,1),\ (5,6)\}$

		State			
		00	01	10	11
	A	1	19	0	0
	B	0	20	0	0
	C	19	0	1	0
	D	20	0	0	0
	E	11	0	8	1
	F	9	7	3	1
	G	0	0	18	2
	H	1	10	5	4
	I	0	18	0	2
Character Class	J	11	0	9	0
	K	6	0	14	0
	L	0	0	1	19
	M	14	6	0	0
	N	19	1	0	0
	O	3	17	0	0
	P	0	1	1	18
	Q	2	2	12	4
	R	0	2	0	18
	S	0	0	18	2
	T	0	0	18	2
TR		1.1542	1.1818	1.3643	1.6229

500×500 to 10×10 sophisticated reduction.   ATR = 1.3013 bits.

We also experiment with forming the n-tuples from a randomly ordered list of points. In Section IV their performance is
compared with that of the nonrandom ones described above.

3.   RECOGNITION SCHEME

The recognition scheme is basically that of Browning and
Bledsoe (1959) also described in detail by Steck (1962) and
Uhr (1973). If an unknown pattern puts the n-tuple into state
j (say), then the jth row of the state matrix for that n-tuple
$(n_{1j},\ n_{2j},\ \ldots,\ n_{sj})$ is used as a score vector (n_{ij} can be

thought of as the score--weight of evidence--of that n-tuple
in favor of X_i). Summing the individual score vectors over
the n-tuples used in the recognition process gives a total
score vector, and the unknown pattern is identified as belonging to the class(es) with highest score(s). We also
experiment with forming the total score vector by a weighted
sum of the individual score vectors where the weights are
various functions of the state transmission rate. In general,
we see that the use of weights is not worth the added effort
except for the randomly chosen n-tuples.

As an example, consider an unknown pattern that puts the
best and second-best "sophisticated" points in state 1. Then
the unweighted sum of the score vector is

(19 20 1 0 10 12 22 23 22 9 14 39 6 1 17 38 22 38 22 22)

and the score vector for the 2-tuple composed of these points is

(0 0 0 0 1 1 2 4 2 0 0 19 0 0 0 18 4 18 2 2)

and we begin to favor the hypothesis that the unknown pattern
is of class L or possibly P or R .

If the vectors are weighted by their TR then the first
would be multiplied by .7691 and the second by .7560 before
adding. The weight associated with the first 2-tuple is
1.6229, which is slightly greater than the sum of the weights
of its constitutents.

We also try using $(n_{1j}/m_j, n_{2j}/m_j, \ldots, n_{sj}/m_j)$ as the
score vector in which the score n_{ij}/m_j is the estimated
conditional probability of class i given the n-tuple state j .

4.   RESULTS

The results are presented in Tables 3 and 4, where the
total number of errors made is tabulated as a function of
various learning and reading situations, and in Table 5 which
tabulates the kind of errors made. Note, too, in Table 5 that

TABLE 3

Number of Errors Made (out of 400) in
"Learn all/Read all" as a Function of
Number of Points Used for Various Recognition Schemes

WEIGHT = 1

No. Pts. Used	1 - tuples						2 - tuples					
	n_{ij}			n_{ij}/m_j			n_{ij}			n_{ij}/m_j		
	H	E	RE	H	E	RE	H	E	RE	H	E	RE
10	111	107	189	112	109	198	118	121	189	127	128	193
20	44	41	87	59	40	92	53	39	86	60	46	83
30	26	30	50	27	22	50	25	28	50	22	29	51
40	18	18	28	22	21	28	17	21	34	15	22	33
50	16	18	25	22	19	21	17	21	34	13	20	25
60	16	12	25	18	11	18	15	13	23	11	12	17
70	14	14	13	18	11	12	9	16	11	11	14	13
80	9	10	10	17	12	12	5	10	8	9	11	10
90	8	8	11	13	10	11	7	7	10	9	10	9
100	8	7	7	13	9	9	6	8	9	10	10	8

No. Pts. Used	5 - tuples					
10	115	118	193	83	104	160
20	61	58	101	40	29	80
30	34	35	62	18	19	46
40	22	18	28	9	11	23
50	17	17	24	7	8	13
60	15	10	21	4	4	11
70	13	10	13	4	4	7
80	10	9	7	4	4	3
90	10	4	12	4	3	1
100	7	7	7	3	3	0

H denotes sophisticated reduction from 500×500 to 10×10 array;
E denotes simple-minded reduction from 500×500 to 10×10 array;
RE denotes random choice of n-tuples for simple-minded reduction.

Table 3 continued

$$\text{WEIGHT} = 2^{TR}$$

| No.
Pts.
Used | 1 - tuples | | | | | | 2 - tuples | | | | | |
| | n_{ij} | | | n_{ij}/m_j | | | n_{ij} | | | n_{ij}/m_j | | |
	H	E	RE	H	E	RE	H	E	RE	H	E	RE
10	110	108	186	134	125	216	114	127	191	135	139	208
20	54	34	85	65	49	121	53	39	87	75	65	121
30	26	24	41	37	21	78	23	24	52	29	48	84
40	22	18	25	26	24	63	15	19	30	23	30	66
50	19	19	23	30	21	47	15	20	22	19	24	46
60	18	11	23	26	16	32	11	13	19	14	20	33
70	17	10	11	24	13	28	10	12	15	13	18	27
80	13	10	6	19	12	24	9	11	9	15	18	21
90	12	8	9	20	13	22	7	10	10	15	13	18
100	11	8	8	20	19	19	8	9	8	14	13	16

	5 - tuples					
10	125	119	165	94	117	175
20	65	65	91	49	56	99
30	35	32	64	32	29	72
40	25	22	32	15	15	46
50	17	16	27	11	13	30
60	12	12	17	9	6	19
70	10	10	11	7	6	14
80	7	9	8	5	6	8
90	6	6	9	6	4	5
100	6	5	5	5	4	3

H denotes sophisticated reduction from 500×500 to 10×10 array;
E denotes simple-minded reduction from 500×500 to 10×10 array;
RE denotes random choice of n-tuples for simple-minded reduction.

TABLE 4

Number of Errors Made (out of 200) in
"Learn last 10/Read first 10" as a Function of
Number of Points Used for Various Recognition Schemes

WEIGHT = 1

No. Pts. Used	1 - tuples						2 - tuples					
	n_{ij}			n_{ij}/m_j			n_{ij}			n_{ij}/m_j		
	H	E	RE	H	E	RE	H	E	RE	H	E	RE
10	73	83	117	75	77	124	79	84	119	85	85	133
20	41	44	71	46	41	75	50	47	77	55	45	81
30	34	39	49	38	31	48	37	36	53	43	36	53
40	29	25	39	27	26	38	29	29	39	34	26	45
50	22	23	32	27	22	35	31	24	33	30	22	36
60	20	18	31	25	13	33	26	18	28	28	17	32
70	19	14	20	21	12	24	25	18	16	23	18	24
80	18	17	19	17	11	22	21	15	18	23	15	20
90	20	20	19	21	15	19	22	18	21	24	16	20
100	20	24	24	19	17	17	22	21	20	24	16	20

	5 - tuples					
10	78	89	138	88	87	127
20	65	65	95	57	60	87
30	48	47	72	49	50	64
40	37	33	55	38	41	50
50	32	33	44	33	38	39
60	27	24	38	26	24	37
70	28	23	31	26	28	30
80	24	21	27	27	26	29
90	25	27	25	28	31	24
100	31	30	22	28	30	21

H denotes sophisticated reduction from 500×500 to 10×10 array;
E denotes simple-minded reduction from 500×500 to 10×10 array;
RE denotes random choice of n-tuples for simple-minded reduction.

Table 4 continued

$$\text{WEIGHT} = 2^{TR}$$

No. Pts. Used	1 - tuples						2 - tuples					
	n_{ij}			n_{ij}/m_j			n_{ij}			n_{ij}/m_j		
	H	E	RE	H	E	RE	H	E	RE	H	E	RE
10	73	79	116	91	76	133	89	83	122	92	89	136
20	44	43	73	57	41	84	54	45	77	75	55	89
30	37	35	47	37	35	61	41	33	53	62	52	77
40	28	24	42	29	32	55	31	24	42	51	39	66
50	24	21	32	32	23	48	29	20	35	46	33	59
60	23	16	34	24	16	47	24	15	31	41	25	56
70	19	14	21	22	18	38	21	15	22	38	22	43
80	16	13	19	18	18	35	20	14	18	30	20	40
90	19	13	15	22	24	35	21	18	19	30	31	35
100	18	16	16	36	35	35	21	18	18	34	36	34

	5 - tuples					
10	81	99	129	93	88	128
20	71	53	94	67	70	98
30	58	47	68	72	69	88
40	48	37	50	62	70	75
50	42	33	46	55	65	64
60	36	26	42	46	58	60
70	32	26	29	42	53	57
80	30	21	26	43	51	54
90	27	22	23	44	50	44
100	26	22	19	44	51	42

H denotes sophisticated reduction from 500×500 to 10×10 array;
E denotes simple-minded reduction from 500×500 to 10×10 array;
RE denotes random choice of n-tuples for simple-minded reduction.

TABLE 5

Kinds of Errors Made for
"Learn last 10/Read first 10" for the Simple-Minded
Reduction, Weight = 1, and Score Vector (n_{ij}) When
the First 50 1-tuples are Used

Character Was	Character Read As																			
	A	B	C	D	E	F	G	H	I	J	K	L	M	N	O	P	Q	R	S	T
A	10																			
B	1	9																		
C			6	1						1				2						
D				10																
E				1	9															
F						10														
G							10													
H								10												
I									10											
J										10										
K											10									
L												10								
M													10							
N													1	9						
O		1		1				1							7					
P																9		1		
Q																	10			
R		1														6		3		
S							2												5	3
T																				10

even though only 50 points are used and 23 errors made there
are 12 pattern classes in which no errors are made and 4 others
in which only one error is made.

5. CONCLUSIONS

I address myself principally to conclusions to be drawn
from Table 4 where the training and recognition sets are disjoint. While a recognition scheme can, hopefully, recognize
what it has learned, the "proof of the pudding" is in recogniz-
ing new instances of patterns.

1. The principal conclusion is that my modification of the Browning-Bledsoe recognition scheme pays off only if not all the points are used. In particular, something less than 70-80% should be used.

2. Another important conclusion is that information weighting does not appear to be of particular value.

3. Using conditional score vectors (n_{ij}/m_j) may be of value when 1-tuples are used.

4. The sophisticated reduction from the large mosaic to the small is not any good at all. This must be because the optimization of the cut-off for a point is plagued by local maxima.

5. I think that the use of informative points would be of great value in a sequential recognition scheme where separation of "look-alikes" is saved until last.

6. Use of n-tuples, $n > 1$, is not justified by these results. It is to be expected, however, that a suitable ordering of n-tuples by their transmission rates would make their use much more attractive.

BIBLIOGRAPHY

Bledsoe, W. W. and Browning, I.   (1960). Pattern recognition and reading by machine. *Proc. Fall Joint Comp. Conf., 1959,* 225.

Steck, G. P.   (1962). Stochastic model for the Browning-Bledsoe pattern recognition scheme. *IRE Trans. Elect. Comp.* <u>EC-11</u>, 274.

Uhr, L. (1973). *Pattern Recognition-Learning and Thought-Computer-Programmed Models of Higher Mental Processes.* New Jersey: Prentice-Hall.

<u>Supplemental Listings</u>

Gilbert, E. S. (1968). On discrimination using qualitative variables. *J. Amer. Statist. Assoc.* <u>63</u>, 1399.
Treats only two class of patterns.

Casey, R. and Nagy, G.   (1966).   Recognition of printed
     chinese characters.   *IEEE Trans. Elect. Comp.* EC-15, 91.
     Uses twenty copies of each of 1000 characters and sequen-
     tial masking.

THE ROLE OF CANONICAL VARIABLES IN DISCRIMINANT ANALYSIS

Anant M. Kshirsagar

Institute of Statistics
Texas A&M University
College Station, Texas

1. INTRODUCTION

In general, canonical analysis and in particular, factorization of Wilks' Λ criterion (corresponding to the direction and collinearity aspects of an assigned function) have been exploited somewhat less often than they could have been. One still finds statisticians contenting themselves with an overall test of significance instead of exploring the structure of the relation in greater detail. Multivariate methods are disliked by many because of the heavy distribution theory involved. Yet, if one reviews the research in multivariate analysis in this country (and elsewhere too) during the last twenty years or so, one finds that the emphasis is on the derivations of the distributions (central and noncentral) of the individual canonical correlations, their moments, and moments of their functions, etc. A very large number of papers have been published in this area. The results are invariably in terms of infinite series, Bessel functions with matrix arguments, zonal polynomials, Meijer's G-function and Mellin-Barnes type integrals. Almost none of these results are useful in any practical problems. So, when it comes to concrete analysis of any multivariate data, the statistician has nothing to offer except Roy's largest root, Hotelling's generalized T^2, Pillai's criterion or Wilks' Λ criterion. These are all overall criteria, and do not tell

us much except that the "groups" are significantly different,
which the experimenter probably knows already. Some research
has already been done about the comparative merits and de-
merits of these criteria but this also has not yielded any
fruitful information. A vague general conclusion is usually
reached, that as far as "power" is concerned, all these criteria
are equivalent. However this is not so. Wilks' Λ criterion
has far more potential than the other criteria. It contains
a large amount of information about the structure of the
groups. Exploitation of this, using the direction and collinearity factors (first derived by Williams (1952a) and
Bartlett (1951)), is still unknown to many practical statisticians. This use of Wilks' Λ criterion has not received much
attention in spite of the excellent expository article by
Williams (1967) in the journal of the Royal Statistical Society.
The distribution theory required for this is very modest, does
not involve complicated mathematical functions and the resulting statistical tests are very simple. Even so, they have
not been given proper attention. Therefore, I thought it
would be of some help to practical workers to reiterate and
review the usefulness of canonical variables and canonical
correlations, in general, and of Wilks' Λ and its factors, in
particular, in the field of discriminant analysis and pattern
recognition.

R. A. Fisher was first to identify the similarity between
regression analysis and discriminant analysis in the case of
two groups. He did so by defining a dummy variable taking
values $n_2/(n_1+n_2)$ and $-n_1/(n_1+n_2)$ corresponding to the
two groups. He then utilized this approach to derive an exact
test of goodness of fit of a hypothetical discriminant
function for two groups. Williams (1952, 1955) and Bartlett
(1951) then extended Fisher's method to several groups. More
than one dummy variable is needed in this general case, and
then discriminant analysis for several groups simply becomes

canonical analysis between the vector of these dummy variables
and the vector of the actual variables measured. This canonical analysis throws more light on the structure of the groups.
One reason for the comparative obscureness of these tests and
techniques, is that the tests were originally presented in
terms of the canonical variables and correlations rather than
the original variables. This presentation makes the tests
appear more difficult to apply than they actually are. However,
the test statistics have been described later in terms of the
original variables [see Kshirsagar (1964, 1969); Williams
(1955, 1967)].

2. THE TYPE OF INFORMATION CANONICAL ANALYSIS CAN PROVIDE

For the sake of definiteness, let us assume that we have
$q + 1$ groups and p correlated characters $x_1, x_2, \ldots, x_p$
are measured for different numbers of individuals belonging to
these groups. If a multivariate analysis of variance is carried
out on the data, we shall obtain the following table for the
corrected sum of squares (s.s.) and sum of products (s.p.)
matrices:

Source	d.f.	s.s. and s.p. matrix (order $p \times p$)
Between groups	q	B
Within groups	$n-q$	A
Total	n	A + B

One can define q dummy variables $y_1, y_2, \ldots, y_q$ by

$$y_i = \begin{cases} 1 & \text{if an individual belongs to the i-th group} \\ & (i = 1, 2, \ldots, q) \\ 0 & \text{otherwise} \end{cases} \qquad (2.1)$$

$$y_{q+1} = 0, \text{ always,} \qquad (2.2)$$

and then B is the matrix of regression s.s. and s.p. in the

regression of the p-vector $\underline{x}$ on the q-vector $\underline{y}$ and A
is the matrix of the error s.s. and s.p. In the Yule-Bartlett
notation B will be $C_{xy} C_{yy}^{-1} C_{yx}$ and A will be

$$C_{xx \cdot y} = C_{xx} - C_{xy} C_{yy}^{-1} C_{yx} .$$

The canonical correlations between $\underline{x}$ and $\underline{y}$ are then
$r_1^2, \ldots, r_k^2,$ where

$$\left| -r_i^2 (A + B) + B \right| = 0 \tag{2.3}$$

and $k = \text{Min}(p, q)$. The canonical variables are $\underline{\ell}_i' \underline{x}$ and

$$[-r_i^2 (A + B) + B] \underline{\ell}_i = 0 \tag{2.4}$$

and $\underline{m}_i' \underline{y}$ is the projection of $\underline{\ell}_i' \underline{x}$ on $\underline{y}$ $(i = 1, 2, \ldots)$.

We shall now briefly describe the type of information that
can be revealed by the r_i^2, $\underline{\ell}_i' \underline{x}$, $\underline{m}_i' \underline{y}$, A and B. For the
actual tests and factors of Wilks' $\Lambda = |A| / |A + B|$ reference
may be made to the papers by Williams and Kshirsagar listed
in the Bibliography or to Kshirsagar (1972).

2.1. Dimensionality of the Configuration

The dimensionality of the configuration of the groups
is estimated from the number of significant canonical correlations from among the r_i^2's. If r_1^2 is the only significant
root of (2.3), the group means are collinear; if r_1^2 and
r_2^2 are both significant, the group means are coplanar and
so on. The number of discriminant functions adequate for discrimination is the same as the dimensionality. Thus if r_1^2
is the only significant root, only one discriminant function
is adequate. [See Binet and Watson (1956) for an algebraic
theory of the computing routine.]

In practice, even if the actual dimensionality is higher,
it is often possible to employ only one or two discriminant
functions, ignoring the requirement of others, as an approximation.

2.2. Discriminant Functions

The discriminant functions are the canonical variables $\underline{\ell}_1'\underline{x}$, $\underline{\ell}_2'\underline{x}$, ..., etc., corresponding to the significant canonical correlations r_1^2, r_2^2, ..., etc. Transforming from the x_1, x_2, ..., x_p axes to new axes, represented by $\underline{\ell}_1'\underline{x}$, $\underline{\ell}_2'\underline{x}$, ..., and ignoring the axes corresponding to the insignificant r_i^2's, and then plotting the group means on this new reference set, gives a sharper, clearer picture of the group configuration. Any new observation is readily converted, in terms of $\underline{\ell}_1'\underline{x}$, $\underline{\ell}_2'\underline{x}$, ..., and plotted on this new graph and then is assigned to that group to which it is the nearest. Thus, this provides a method discrimination and allocation of a new observation to one of the groups.

2.3. Discriminant Functions in the Dummy Variables Space

The canonical variables $\underline{m}_1'\underline{y}$, $\underline{m}_2'\underline{y}$, ..., corresponding to the significant roots are also useful. The elements of $\underline{m}_1$ provide the optimum scores for the groups. The relative importance of the different groups is therefore brought out by the vector $\underline{m}_1$. Further $\underline{m}_1'\underline{y}$, $\underline{m}_2'\underline{y}$, ..., correspond to contrasts among the groups. So, this method not only tests the overall significance of the differences among the groups, but picks out those contrasts that are significant and those that are not. Therefore, it is not only able to declare the existence of differences among the group means, but also the magnitude and nature of these differences.

2.4. Selection of Variables

By factorizing Λ as

$$\Lambda_{x_1} \; \Lambda_{x_2|x_1} \; \Lambda_{x_3|x_1,x_2} \; \cdots$$

and testing the individual factors, or by considering Λ based on $\underline{x}$ and Λ based only on a subset of $\underline{x}$, one can determine which variables are important for discrimination and which are not. Those that do not contribute sufficiently to Λ can be dropped without affecting the performance of the

discriminant functions. This is similar to choice of variables
in a multiple regression. [See Eisenbeis and Gilbert (1973)
for more details.]

2.5. Hypothetical Discriminant Function

The experimenter often has one or more linear functions
in mind which he thinks will be good discriminators; e.g.,
the total number of marks in different subjects, i.e.,
$x_1 + \ldots + x_p$, or the total yield $x_1 + x_2$, of straw and
grain or some such functions, which have some physical meaning
also. The question arises of whether these functions are in
agreement with the discriminant functions $\underline{\ell}_1'\underline{x}$, $\underline{\ell}_2'\underline{x}$, ...,
etc. derived from the data. If they are, these functions
naturally are to be preferred as they are simpler and are
likely to have a physical interpretation. Williams (1952) and
Bartlett (1951) propose a test for this by factorizing Wilks'
Λ as

$$\Lambda_H \, \Lambda_R$$

where Λ_H is Λ based on the assigned or proposed discriminant functions, and Λ_R is the "residual", when the
proposed functions are eliminated. Then, they further factor-
ize Λ_R as $\Lambda_D \Lambda_{C|D}$ or $\Lambda_C \Lambda_{D|C}$. The proposed discriminant
functions may not be suitable because (a) their number is not
adequate <u>and/or</u> (b) they do not agree in their directions with
the "true" discriminant functions. The dimensionality aspect,
(a), and the direction aspect, (b), of the hypothesis of goodness of fit of the proposed discriminant functions are tested
by Λ_D and $\Lambda_{C|D}$ or by Λ_C and $\Lambda_{D|C}$, which is an alter-
native factorization. Here, Λ_D is the "total" direction
factor and $\Lambda_{C|D}$ is the "partial" or conditional dimensionality
factor. Similarly, Λ_C is the "total" dimensionality factor
and $\Lambda_{D|C}$ is the "partial" or conditional direction factor.
These two factorizations are analogous to the factors $p(A)$,
$p(B|A)$ or $p(B)$, $p(A|B)$ of the total probability $p(AB)$ of
two events, A and B.

2.6. Hypothetical Discriminant Functions in Dummy Variables Space

It is possible to consider a similar factorization of Wilks' Λ when the proposed discriminant functions are given in terms of the dummy y variables. Theoretically this presents no new problems, because in canonical analysis, the $\underline{x}$ and $\underline{y}$ variables are symmetric with respect to each other, and Wilks' Λ, which is $|C_{xx \cdot y}|/|C_{xx}|$, is invariant under an interchange of x and y. So the direction and dimensionality factors in this case can be obtained theoretically by simply interchanging x and y. The problem in this case is to express these factors in terms of the matrices A, B of the analysis of variance table and the proposed functions. This has been achieved and hence it is not essential in any problem to bring out the y variables explicitly and formulate C_{yy}, C_{yx}, etc. The null distributions of test statistics remain the same whether $\underline{x}$ is normal and $\underline{y}$ is dummy or vice versa.

2.7.   Contingency Table

Sometimes, like the y variables, the x variables are also dummy. This can happen if we are considering a contingency table for two attributes. The x variables correspond to the categories of one attribute and the y variables correspond to the categories of the other attribute. The matrices A and B of the analysis of variance table then are obtained in terms of the frequencies n_{ij} of the contingency table. The number of significant canonical correlations then tells not only whether there is a relationship between the two attributes but also whether that relationship is "linear" or "quadratic", etc. The elements of $\underline{\ell}_1$ in $\underline{\ell}_1'\underline{x}$ and of $\underline{m}_1$ in $\underline{m}_1'\underline{y}$ provide a "quantification" of the qualitative categories of the attributes. Thus the categories which were simply A_1, A_2, A_3, ... to the experimenter are no longer so after the canonical analysis is performed. One can reach conclusions such as A_2 is nearer to A_1 than A_3, and that A_5 is the lowest and A_7 is the highest, and so on.

2.8. Null Functions

So far we have discussed the significant canonical correlations only and the corresponding canonical variables. But
even the insignificant roots of (2.3) and the corresponding
canonical variables are useful in certain situations. This
does not arise in discriminant analysis or pattern recognition
but is mentioned here to complete the account of usefulness of
canonical variables. When $\underline{x}$ and $\underline{y}$ are both economic
variables (including indogeneous and exogeneous variables), the
canonical variables corresponding to the insignificant roots
yield, under suitable conditions, structural relations for the
economy. By suitably combining these canonical variables and
by employing the "identification" conditions of structural relations, estimates of the structural relations can be derived.
[See Anderson and Rubin (1949).]

3. SOME ILLUSTRATIVE EXAMPLES

Fisher (1946) and later Bartlett (1951) give a complete
analysis of Taylor's blood serological data where the blood
reactions are described by the symbols -, ?, w, (+) and +.
Canonical analysis is employed to see whether these symbols
can be represented on a linear scale. It is found that this is
possible and the optimum scores for these "groups" or symbols
turn out to be

    -     0
    ?   0.19
    w   0.58
    (+)   0.96
    +   1.00

Factorization of Wilks' Λ criterion to test the goodness of
fit of hypothetical scores 0, 1/4, 1/2, 3/4, 1 to -, ?, w,
(t), +, respectively, is described by Bartlett. The conclusion
is that the hypothetical scores are not in agreement with the

derived ones, showing that the symbols are not equi-spaced on the linear scale, the serious departure being due to (+).

Williams (1967) illustrates the technique of canonical analysis on certain characteristics of lamb carcasses. In this case the groups are six and are formed by the combination of two weight classes and three grades. The dimensionality is found to be two. The canonical analysis shows that the main-effects of the two factors' weight and grade are significant while their interaction is not. Factorization of Wilks' Λ is employed to study the adequacy of two simple hypothetical discriminant functions in the y-space. Equally spaced scores for the three grades are also tested and found to be inadequate, the optimum scores being 1, 5 and 6 (which are certainly different from equi-spaced scores).

Data requiring two or more discriminators are difficult to interpret, but if the derived discriminators or the proposed discriminators (in case they are in agreement with the derived ones) have some physical meaning, such data can yield a satisfactory interpretation.

M. M. Barnard's (1935) data on four series of Egyptian skulls analyzed by Rao (1955), Bartlett (1947), Williams (1959), and Kshirsagar (1962), is also a very interesting illustration of the use of canonical analysis. The proposed discriminator in this case is "time" and its values corresponding to the four series are given. Canonical analysis shows that no single discriminator can be adequate for this data. But if departure from collinearity of the four groups is ignored then time can be employed. The proposed discriminator in this example is from the space of the dummy variables. The question is, whether the three degrees of freedom corresponding to the differences among the series are such that only one of them is significant and the other two are not, or, in other words, whether only one contrast among the four series is responsible for the significant group differences. If so, the problem is

to determine this contrast and canonical analysis can resolve
this. [See Kshirsagar (1972).]

Williams (1952b) gives optimum scores for Peridontal conditions by employing a canonical analysis on the data of the
contingency table where the rows refer to the Peridontal conditions and the columns to the amount of calcium intake. He
reaches the conclusion that, above a certain ceiling, increase
in the amount of calcium intake has no effect on the condition.

A very different and interesting application of canonical
variables and correlations is supplied by Patel (1970) who
proceeds to derive optimum scores for the states of a Markov
chain. The variables $N_i(t)$ are defined

$$N_i(t) = \begin{cases} 1 & \text{if the Markov chain is in state i at time t} \\ 0 & \text{otherwise} \quad (i = 1, 2, \ldots, m) \end{cases}$$

$$(3.1)$$

Let $\underline{N}(t)$ be the column vector of these variables. He then
finds the canonical correlations and canonical variables for
the two vector variables $\underline{N}(t)$ and $\underline{N}(t+1)$. Patel shows that
the optimum scores are provided by the eigen-vector of the
transition probability matrix of the Markov chain, corresponding to the largest eigenvalue, ignoring the obvious eigenvalue
1. Kshirsagar and Arseven (1973) extend this result to Markov
Renewal Processes and obtain a similar result. Williams (1952c)
and Kshirsagar (1965) describe the use of canonical analysis
in factorial experiments when the interaction effect between
the two factors is multiplicative.

Green and Halbert (1966) give an interesting application
of canonical analysis in marketing research in attempting to
relate information on buying behavior to various personality
characteristics of subjects participating in the experiment.

Bartlett (1948) demonstrates the use of canonical analysis
in estimating supply and demand relations from time series
data giving observations on consumption, price, and cost of
cotton yarn and the income of consumers. This illustration

shows how even the null functions, i.e., canonical variables corresponding to insignificant canonical correlations, are useful. Unfortunately the factorization of Wilks' Λ criteria to test the goodness of fit of a proposed null function is not possible, though Williams (1955) erroneously presents it so. [See Bartlett (1957) for discussion of the fallacy in Williams' argument.] [Also see Kshirsagar (1962) for the use of null functions in causal chain models.]

Barnett and Lewis (1963) study the relation between G.C.E. and degree results in England by the use of canonical analysis and demonstrate the use of canonical variables for prediction purposes.

4. AREAS OF FURTHER RESEARCH

We consider two situations to illustrate areas of further research for increasing the potential usefulness of canonical analysis.

<u>Situation A.</u>

Suppose both $\underline{x}$ and $\underline{y}$ contain too many variables. This occurs when one is dealing with economic variables or investigating an entirely new, unexplored area. The natural tendency then is to include as many variables as one can think of, caring little for internal functional relations among them or for multicollinearity. Then the matrices C_{xx}, C_{yy} or both become singular and determination of canonical correlations and variables becomes difficult. Possibly the use of principal components analysis on C_{xx} and C_{yy} or the use of generalized inverses of matrices may get rid of the singularities without destroying the symmetry in the formulae. The effect of these singularities on the interpretation and validity of statistical tests of canonical correlations needs to be investigated thoroughly.

It is conjectured that even the formulae for direction and collinearity factors would remain the same algebraically if generalized inverses of C_{xx} and C_{yy} are employed in

place of the ordinary inverses. [This conjecture is now proved
to be correct by Kshirsagar and Wilson (1974), Ph.D. dissertation,
Texas A&M University).] The reason for this conjecture is that
Kshirsagar (1970a) is able to show this while dealing
with the categories of a contingency table, and Patel (1970)
also successfully employs a similar approach in his Markov
chain example, though he does not justify it. However, one
needs to be very careful in such situations to reject certain
irrelevant roots and the corresponding eigenvectors. It is
therefore essential to evolve a systematic theory of this
approach. It is also suspected that the test of the goodness
of fit of any proposed discriminant function would remain the
same, even if the proposed discriminant function, say $\underline{h}'\underline{x}$, is
replaced by its linear combination with those functions $\underline{a}'\underline{x}$
of the variables $\underline{x}$, which create singularities in C_{xx}.
It would thus be difficult to distinguish between $\underline{h}'\underline{x}$
and $\underline{h}'\underline{x} + \lambda\underline{a}'\underline{x}$. In other words any proposed discriminant
function would get hopelessly confounded with singularities
of C_{xx}, and a similar ambiguity would exist for the y-space,
if C_{yy} is also singular. In fact, it is possible that the
goodness of fit of certain $\underline{h}'\underline{x}$'s cannot even be tested,
especially if they are linear combinations of only the null
functions $\underline{a}'\underline{x}$. Criteria of testability of proposed functions
would therefore have to be set up first.
Situation B.

     It is usually assumed in classical canonical analysis that
the same number of observations are available on each character.
But if this is not so, and certain observations on certain
variables are either not recorded or are destroyed, the classical procedure of estimation of canonical correlations and
variables of Section 2 breaks down. Therefore, it is essential
to obtain good and reliable estimates of these quantities
from the partial available data, and the effect of this on the
formulation of Wilks' Λ and its direction and collinearity
factors must be investigated.

5. CONCLUDING REMARKS

Canonical analysis is an extremely useful technique of multivariate analysis. It yields valuable information in a number of ways. The objective of this paper has been to list its advantages for the benefit of practical statisticians. However, it should not be assumed from this discussion that canonical analysis is applicable in each and every situation: canonical analysis has its own limitations and disadvantages. For example, if the variance-covariance matrix of $\underline{x}$ is not the same for all the groups, the use of canonical analysis is not justifiable. Similarly, the propriety of the use of sophisticated tests when neither $\underline{x}$ nor $\underline{y}$ is normal is also questionable, though practical statisticians like Bartlett (1963) advocate it, advising that the tests of significance should not be taken too precisely in such cases.

APPENDIX

A.1. Wilks' Λ Test

If A has the Wishart distribution $W_p(A|n-q|\Sigma)$ [see Kshirsagar (1972) for notation] and B has an independent Wishart distribution $W_p(B|q|\Sigma)$, we say that $\Lambda = |A|/|A+B|$ has Wilks' $\Lambda(n, p, q)$ distribution. If $p = 1, 2$ or $q = 1, 2$ we use the following tests.

$$\text{(i)} \quad \frac{1-\Lambda}{\Lambda} \cdot \frac{n-q}{q} = F_{q,n-q}, \text{ when } p = 1$$

$$\text{(ii)} \quad \frac{1-\Lambda}{\Lambda} \cdot \frac{n-p}{p} = F_{p,n-p}, \text{ when } q = 1$$

$$\text{(iii)} \quad \frac{1-\sqrt{\Lambda}}{\sqrt{\Lambda}} \cdot \frac{n-q-1}{q} = F_{2q,2(n-q-1)}, \text{ when } p = 2$$

$$\text{(iv)} \quad \frac{1-\sqrt{\Lambda}}{\sqrt{\Lambda}} \cdot \frac{n-p-1}{p} = F_{2p,2(n-p-1)}, \text{ when } q = 2$$

Otherwise, we use

$$W = -\left[n - \frac{1}{2}(p+q+1)\right] \log_e \Lambda \qquad\qquad (A.1.1)$$

which is a χ^2 with pq distribution function (d.f.). For more precise results, we find

$$C_\alpha (p, q, M) \qquad\qquad (A.1.2)$$

where α = level of significance, $M = n-p-q+1$, from Schatzoff's (1966) tables and multiply this $C_\alpha(p, q, M)$ by the $100\alpha\%$ point of a χ^2 with pq d.f. This product is the exact $100\alpha\%$ point of W.

A.2. <u>Test of Hypothesis That Certain s Roots Are Significant</u>

The test of the hypothesis that only s roots r_1^2, r_2^2, ..., r_s^2 are significant is

$$-\left\{ n - s - \frac{1}{2}(p+q+1) + \sum_1^s \frac{1}{r_i^2} \right\} \log_e \prod_{i=s+1}^{k} (1 - r_i^2)$$

is a χ^2 with $(p-s)(q-s)$ d.f. This test is sequentially carried out for $s = 1, 2, \ldots$, etc., to determine the number of significant roots.

A.3. Proposed Discriminant Function $\underline{h}'\underline{x}$

$$\Lambda = \frac{|A|}{|A+B|}, \quad \Lambda_H = \frac{\underline{h}'A\underline{h}}{\underline{h}'(A+B)\underline{h}}, \quad \Lambda_R = \frac{\Lambda}{\Lambda_H}$$

<table>
<tr><td align="center"><u>Factor</u></td><td align="center"><u>Distribution</u></td></tr>
<tr><td align="center">$\Lambda_D = \dfrac{1 - \dfrac{\underline{h}'B(A+B)^{-1}B\underline{h}}{\underline{h}'B\underline{h}}}{\Lambda_H}$</td><td align="center">$\Lambda(n-1,\ 1,\ p-1)$</td></tr>
<tr><td align="center">$\Lambda_{C|D} = \Lambda_R / \Lambda_D$</td><td align="center">$\Lambda(n-2,\ q-1,\ p-1)$</td></tr>
</table>

Alternatively

$$\Lambda_C = \Lambda \cdot \left\{ 1 + \frac{\underline{h}'B\,A^{-1}B\underline{h}}{\underline{h}'B\underline{h}} \right\} \qquad \Lambda(n-1,\ q-1,\ p-1)$$

$$\Lambda_{D|C} = \Lambda_R / \Lambda_C \qquad\qquad\qquad \Lambda(n-q,\ 1,\ p-1)$$

A.4. Proposed Discriminant Function $t = \underline{m}'\underline{y}$ in the y-space

The corrected s.s. and s.p. supplied C_{tt} and C_{tx} .

Residual Λ is then $\Lambda_R = |A| / |A+B - C_{xt}C_{tt}^{-1}C_{tx}|$.

Factor		Distribution

$$\Lambda_D = \frac{C_{tx}(A+B)^{-1}A(A+B)^{-1}C_{xt}}{C_{tx}(A+B)^{-1}C_{xt}} \cdot \frac{C_{tt}}{C_{tt}-C_{tx}(A+B)^{-1}C_{xt}} \qquad \Lambda(n-1,\ 1,\ q-1)$$

$$\Lambda_{C|D} = \Lambda_R/\Lambda_D \qquad\qquad\qquad\qquad \Lambda(n-2,\ p-1,\ q-1)$$

Alternatively

$$\Lambda_C = \frac{|A|}{|A+B|} \cdot \frac{C_{tx}A^{-1}C_{xt}}{C_{tx}(A+B)^{-1}C_{xt}} \qquad \Lambda(n-1,\ p-1,\ q-1)$$

$$\Lambda_{D|C} = \Lambda_R/\Lambda_C \qquad\qquad\qquad\qquad \Lambda(n-p,\ 1,\ q-1)$$

A.5. More than One Proposed Discriminant Functions
(from the x-space)

Proposed s functions $\Gamma'\underline{x}$

$$\Lambda_H = |\Gamma'A\Gamma| / |\Gamma'(A+B)\Gamma|$$

$$\Lambda_R = \Lambda/\Lambda_R$$

Factor		Distribution

$$\Lambda_D = \frac{|\Gamma'B(A+B)^{-1}A\Gamma|}{\Lambda_H|\Gamma'B\Gamma|} \qquad\qquad \Lambda(n-s,\ s,\ p-s)$$

$$\Lambda_{C|D} = \Lambda_R/\Lambda_D \qquad\qquad\qquad\qquad \Lambda(n-2s,\ p-s,\ q-s)$$

Alternatively

$$\Lambda_C = \frac{\Lambda|\Gamma'B\Gamma + \Gamma'B\,A^{-1}\,B\Gamma|}{|\Gamma'B\Gamma|} \qquad\qquad \Lambda(n-s,\ q-s,\ p-s)$$

$$\Lambda_{D|C} = \Lambda_R/\Lambda_C \qquad\qquad\qquad\qquad \Lambda(n-q,\ s,\ p-s)$$

A.6. Proposed Discriminant Functions (s in number) in the y-space $\underline{t}$

Given quantities C_{tt}, C_{tx}, let

$$C_{tt \cdot x} = C_{tt} - C_{tx} C_{xx}^{-1} C_{xt}$$

$$= C_{tt} - C_{tx} (A+B)^{-1} C_{xt}$$

Then

$$\Lambda = \frac{|A|}{|A+B|} \;, \quad \Lambda_R = \Lambda \cdot \frac{|C_{tt}|}{|C_{tt \cdot x}|}$$

Factor	Distribution
$\Lambda_D = \dfrac{\left|C_{tx}(A+B)^{-1} A (A+B)^{-1} C_{xt}\right|\,\left|C_{tt}\right|}{\left|C_{tx}(A+B)^{-1} C_{xt}\right|\,\left|C_{tt \cdot x}\right|}$	$\Lambda(n-s,\ s,\ q-s)$
$\Lambda_{C|D} = \Lambda_R / \Lambda_D$	$\Lambda(n-2s,\ p-s,\ q-s)$

Alternatively

$$\Lambda_C = \Lambda \cdot \frac{\left|C_{tx} A^{-1} C_{xt}\right|}{\left|C_{tx}(A+B)^{-1} C_{xt}\right|} \qquad\qquad \Lambda(n-s,\ p-s,\ q-s)$$

$$\Lambda_{D|C} = \Lambda_R / \Lambda_C \qquad\qquad\qquad\qquad \Lambda(n-p,\ s,\ q-s)$$

A.7. The Hypothesis of Independence of X_i with $\underline{y}$

Let A_i, $A_i + B_i$ denote matrices obtained from the first i rows and i columns of A and $A+B$ respectively. Then the hypothesis of independence of x_i with $\underline{y}$, where $x_1, \ldots, x_{i-1}$ are eliminated (i.e., the hypothesis of use-lessness of x_i, in presence of $x_1, \ldots, x_{i-1}$) is tested by

$$\Lambda_{i \cdot 12 \ldots i-1} = \frac{\Lambda_i}{\Lambda_{i-1}} = \frac{|A_i|/|A_i + B_i|}{|A_{i-1}|/|A_{i-1} + B_{i-1}|}$$

and has the $\Lambda(n-(i-1),\ 1,\ q)$ distribution.

A.8. The Hypothesis of Independence of $[x_{k+1}, \ldots, x_p]$ with $\underline{y}$

Independence of $[x_{k+1}, \ldots, x_p]$ with $\underline{y}$ when $[x_1, \ldots, x_k]$ are eliminated (i.e., hypothesis of no use of $x_{k+1}, \ldots, x_p$ in discrimination when $x_1, \ldots, x_k$ are present) is tested by

$$\Lambda_{k+1, \ldots, p \cdot 12 \ldots k} = \frac{|A|/|A + B|}{|A_k|/|A_k + B_k|}$$

and has the $\Lambda(n-k, p-k, q)$ distribution.

A.9. Discriminating Procedure

Significant canonical variables are $u_1 = \underline{\ell}_1'\underline{x}, \ldots, u_s = \underline{\ell}_s'\underline{x}$.

Observation: $\underline{x}_o$

Observed values of canonical variables are $u_{01} = \underline{\ell}_1'\underline{x}_0$, $\ldots, u_{0s} = \underline{\ell}_s'\underline{x}_0$. New co-ordinates of the mean of α-th group

$$u_{\alpha 1} = \underline{\ell}_1'\underline{\bar{x}}_\alpha, \ldots, u_{\alpha s} = \underline{\ell}_s'\underline{\bar{x}}_\alpha$$

distance between α^{th} group and the observation, in the s-flat is

$$\Delta_{\alpha 0}^2 = \sum_{i=1}^{s} (u_{\alpha i} - u_{0i})^2$$

Assign $\underline{x}_o$ to π_a if

$$\Delta_a^2 = \text{Min}(\Delta_{10}^2, \ldots, \Delta_{k0}^2)$$

Note: $\underline{\ell}'\underline{x}$ is so determined that

$$[-r_i^2 C_{xx} + C_{xy}C_{yy}^{-1}C_{yx}]\underline{\ell}_i = 0$$

and

$$\underline{\ell}_i'C_{xx}\underline{\ell}_i = 1 \text{ (Normalization)}$$

<u>A.10. Basic Factorization (ξ = hypothetical f^n(x-space))</u>

$$\Lambda = \Lambda_H \Lambda_R$$

$$\iff \frac{|C_{xx \cdot y}|}{|C_{xx}|} = \frac{|C_{\xi\xi \cdot y}|}{|C_{\xi\xi}|} \cdot \frac{|C_{zz \cdot y\xi}|}{|C_{zz \cdot \xi}|}$$

where $\underline{z}$ are p-1 x-variables $\perp$ to ξ linear combinations.
Also, equivalently,

$$\Lambda_H \cdot \Lambda_R$$

$$\frac{|C_{yy \cdot x}|}{|C_{yy}|} = \frac{|C_{yy \cdot \xi}|}{|C_{yy}|} \cdot \frac{|C_{yy \cdot x}|}{|C_{yy \cdot \xi}|}$$

And

$$\Lambda_R \text{ or } \frac{|C_{yy \cdot x}|}{|C_{yy \cdot \xi}|} = \Lambda_D \Lambda_{C|D}$$

$$= \frac{|C_{\tau\tau \cdot x}|}{|C_{\tau\tau \cdot \xi}|} \cdot \frac{|C_{\gamma\gamma \cdot xy}|}{|C_{\gamma\gamma \cdot \xi y}|}$$

where

τ = sample projection of ξ on y-space

$\underline{\gamma}$ = q-1 linear combination of $\underline{y}$'s $\perp$ to τ

Alternatively

$$\Lambda_R = \Lambda_C \Lambda_{D|C}$$

$$= \frac{|C_{\gamma\gamma \cdot x}|}{|C_{\gamma\gamma \cdot \xi}|} \cdot \frac{|C_{\tau\tau \cdot x\gamma}|}{|C_{\tau\tau \cdot \xi\gamma}|}$$

Distributions

$$\Lambda_H \cdot \Lambda_R$$

$$\Lambda(n,p,q) = \Lambda(n,1,q) \; \Lambda(n-1,\, p-1,\, q)$$

$$\Lambda(n-1,\, p-1,\, q) = \begin{cases} \Lambda(n-1,\, p-1,\, 1) \; \Lambda(n-2,\, p-1,\, q-1) \\ \Lambda(n-1,\, p-1,\, q-1) \cdot \Lambda(n-q,\, p-1,\, 1) \end{cases}$$

Interchange $\underline{x}$ with $\underline{y}$, p with q and we have results for a hypothetical function in the y-space.

BIBLIOGRAPHY

Anderson, T. W., and H. Rubin (1949). Estimation of the parameters of a single equation in a complete system of stochastic equations. *Ann. Math. Statist.* <u>20</u>, 46.

Barnard, M. M. (1935). The secular variations of skull characters in four series of Egyptian skulls. <u>*Ann.*</u> *Eugenics* <u>6</u>, 352.

Barnet, V. D., and T. Lewis (1963). A study of the relation between G.C.E. and degree results. *J. R. Statist. Soc. A* <u>126</u>, 187.

Bartlett, M. S. (1947). Multivariate analysis, suppl. *J. R. Statist. Soc.* <u>9</u>, 16.

Bartlett, M. S. (1948). A note on the statistical estimation of supply and demand relations from time series. *Econometrica* <u>16</u>, 323.

Bartlett, M. S. (1951). The goodness of fit of a single hypothetical discriminant function in the case of several groups. *Ann. Eugenics* <u>22</u>, 199.

Bartlett, M. S. (1957). A note on tests of significance for linear functional relationships. *Biometrika* <u>44</u>, 268.

Bartlett, M. S. (1963). "Multivariate Statistics", <u>Theoretical and Mathematical Biology</u>. New York: Blaisdell.

Binet, F. E., and G. S. Watson (1956). Algebraic theory of the computing routine for tests of

significance on the dimensionality of normal
multivariate systems. *J. R. Statist. Soc. B*
18, 70.

Eisenbeis, R. A., and G. G. Gilbert (1973). Investigating
the relative importance of individual variables and
variable subsets in discriminant analysis. *Comm.
Statist. 2*, 205.

Green, P., and M. Halbert (1966). Canonical analysis:
an exposition and illustrative application. *J.
Marketing 3*, 32.

Kshirsagar, A. M. (1962). A note on direction and
collinearity factors in canonical analysis.
Biometrika 49, 255.

Kshirsagar, A. M. (1962). Prediction from the
simultaneous equations system and Wold's im-
plicit causal chain model. *Econometrica 30*, 801.

Kshirsagar, A. M. (1963). Confidence intervals for
discriminant function coefficients. *J. Ind.
Statist. Assoc. 1*, 1.

Kshirsagar, A. M. (1964a). "Wilks' Λ". *J. Ind.
Statist. Assoc. 2*, 1.

Kshirsagar, A. M. (1964b). Distributions of the
direction and collinearity factors in discriminant analysis. *Proc. Camb. Phil. Soc.
60*, 217.

Kshirsagar, A. M. (1965). A note on the use of
canonical analysis in factorial experiments.
J. Ind. Statist. Assoc. 3, 165.

Kshirsagar, A. M. (1969). Distributions associated
with Wilks' Λ. *J. Austral. Math. Soc. 10*, 269.

Kshirsagar, A. M. (1970a). Goodness of fit of an
assigned set of scores for the analysis of

association in a contingency table. *Ann. Inst. Statist. Math.* <u>22</u>, 295.

Kshirsagar, A. M. (1970b). An alternative derivation of the direction and collinearity statistics in discriminant analysis. *Calcutta Statist. Assoc. Bull.* <u>19</u>, 123.

Kshirsagar, A. M. (1971). Goodness of fit of a discriminant function from the vector space of dummy variables. *J. R. Statist. Soc.* B <u>33</u>, 111.

Kshirsagar, A. M. (1972). <u>Multivariate Analysis</u>. New York: Marcel Dekker, Inc.

Kshirsagar, A. M., and E. Arseven (1973). Optimum scores for the states of a Markov renewal process. Submitted for publication.

Patel, H. I. (1970). Choice of optimum scores in a Markov chain of order 1. *Ann. Inst. Statist. Math.* <u>22</u>, 535.

Rao, C. R. (1965). <u>Linear Statistical Inference and its Applications</u>. New York: John Wiley and Sons.

Schatzoff, M. (1966). Exact distribution of Wilks' likelihood ratio criterion. *Biometrika* <u>53</u>, 347.

Srikantan, K. S. (1970). Canonical association between nominal measurements. *J. Amer. Statist. Assoc.* <u>65</u>, 284.

Williams, E. J. (1952a). Some exact tests in multivariate analysis. *Biometrika* <u>39</u>, 17.

Williams, E. J. (1952b). Use of scores for the analysis of association in contingency tables. *Biometrika* <u>39</u>, 274.

Williams, E. J. (1952c). The interpretation of interactions in factorial experiments. *Biometrika* <u>39</u>, 65.

Williams, E. J.   (1955).   Significance tests for dis-
    criminant functions and linear functional relation-
    ships. *Biometrika* 42, 360.

Williams, E. J.   (1959).   Regression Analysis. New
    York: John Wiley and Sons.

Williams, E. J.   (1961).   Tests for discriminant
    functions. *J. Austral. Math. Soc.* 2, 243.

Williams, E. J.   (1967).   The analysis of association
    among many variates. *J. R. Statist. Soc.* B 29,
    199.

AUTOREGRESSIVE, MAXIMUM ENTROPY,
AND G-SPECTRAL ESTIMATION

Henry L. Gray

Southern Methodist University
Dallas, Texas

Manus R. Foster

Mobil Research and Development Corporation
Dallas, Texas

1. INTRODUCTION

In this paper the maximum entropy spectral estimator suggested by Burg (1967) and the autoregressive spectral estimator suggested by Steiglitz (1964) and later by Parzen (1968) are discussed and the extent to which they are equivalent is pointed out. In addition to these estimators a new spectral estimator, referred to as the G-spectral estimator, is introduced which has a certain heuristic appeal, especially if the appropriate model is not definitely known to be a finite autoregressive process.

2. THE AUTOREGRESSIVE MODEL

The discrete autoregressive model is defined as follows. Let $\{Z(t),\ t = 0,\ \pm 1,\ \ldots\}$ be a zero mean white noise process with variance σ_Z^2 and let $\{X(t)\}$ be a stochastic process such that

$$X(t) + a_1 X(t - 1) + \ldots + a_n X(t - n) = Z(t) \qquad (2.1)$$

and

$$E[X(t)Z(S)] = 0, \quad t < S \qquad (2.2)$$

Then we call $\{X(t)\}$ a discrete nth order autoregressive
process. The model defined by (2.1) is a stochastic difference
equation. It can be interpreted as the response of a system
with finite memory to random shocks. On the other hand, as we
shall see, it can also be thought of as an input to a *whitening*
filter. This latter observation will become clear in Section 3.

Now if all of the zeros of

$$p(r) = 1 + a_1 r + a_2 r^2 + \ldots + a_n r^n \tag{2.3}$$

lie outside the unit circle, $\{X(t)\}$ is stationary and stable.
The latter means that $\{X(t)\}$ has a finite variance and can be
written as an infinite moving average; i.e.,

$$X(t) = Z(t) + \sum_{k=1}^{\infty} b_k Z(t - k) \tag{2.4}$$

We shall henceforth tacitly assume the roots of (2.3) lie outside the unit circle when referring to autoregressive processes.
Later we will return to this same condition on $p(r)$ in the
context of the so-called *prediction error filter* (PEF) and in
that case we will refer to it as the minimum phase requirement.

Since by (2.4) $X(t)$ can be expressed as a linear trans-
formation of white noise, it follows that the power spectrum of
$\{X(t)\}$, $S_X(f)$, is given by

$$S_X(f) = \sigma_Z^2 \left| 1 + \sum_{k=1}^{n} a_k e^{-i2\pi fk} \right|^{-2} \tag{2.5}$$

Now from (2.2) it is easily seen that the autocorrelation
$R_X(k)$ satisfies the system of equations

$$R_X(0) + a_1 R_X(1) + \ldots + a_n R_X(n) = \sigma_Z^2 \tag{2.6a}$$

$$R_X(k) + a_1 R_X(k - 1) + \ldots + a_n R_X(k - n) = 0, \quad k \geq 1 \tag{2.6b}$$

where $R_X(k) = R_X(-k)$. It is clear therefore that if estimates
of the autocorrelation $R_X(k)$ are available for each k, then
estimates for the a_k in (2.5) can be obtained through (2.6a,b).
These estimates, when substituted in (2.5), define an estimator
$\hat{S}(f)$ which is called the *autoregressive spectral estimator*; i.e.,

$$\hat{S}_X(f) = \hat{\sigma}_Z^2 \left| 1 + \sum_{k=1}^{n} \hat{a}_k e^{-i2\pi fk} \right|^{-2} \tag{2.7}$$

The commonly used estimators for the $R_X(k)$ from a sample
of size N are of the form

$$\hat{R}_X(k) = W_M(k) \frac{1}{N} \sum_{j=0}^{N-k} X(j)X(j + k) \tag{2.8}$$

where $W_M(k)$ is called a *lag window*, and $M < N$ is the *half-
window length*. Many different windows have been proposed in the
literature (Parzen (1969)) and their statistical properties
investigated. Since multiplication in the time domain goes over
into convolution in the frequency domain, the effect of the lag
window is a smoothing of the estimated spectrum. If $W_M(k)$ is
zero for $|k| > M$, then it is well known that the choice $W_M(k)$
$= 1$ for $|k| \leq M$ gives the least smoothing of the spectrum.
This is called the *rectangular window* and does not necessarily
lead to positive definite estimates. Various authors, Akaike
(1970), Kromer (1969), Parzen (1969) and Steiglitz (1964), have
considered $\hat{S}(f)$ in (2.7) for the rectangular window. Although
considerable theoretical properties have been developed in this
case, criticisms still exist (see Burg (1975) or Koopmans (1974)).
Some other windows do lead to positive definite estimates, but
the situation is not satisfactory theoretically, being liberally
infused with practical art.

In the following sections we will discuss the Burg algorithm,
which provides a practical means of estimating the a_i directly

from the data without the intermediate step of estimating
the correlations $R_X(k)$.

3.   THE MAXIMUM ENTROPY POWER SPECTRUM

The maximum entropy approach to obtaining (2.5) as a suitable model for the power spectrum of a stochastic process, suggested by Burg in 1967, has been discussed by many authors since
that time (e.g., see Ulrych (1972)). Burg himself has never
published his results in this area although he has recently doc-
umented them in a thesis. In this section we sketch the ideas
leading to (2.5) from entropy considerations and then describe
the algorithm suggested by Burg for estimating the a_k.

The entropy of a Gaussian band-limited process has been
shown by Shannon and Weaver (1949) to be proportional to

$$\int_{-w}^{w} \log S(f)\,df \tag{3.1}$$

where $S(f)$ is the power spectrum of the process assumed to
have its total power concentrated in the interval $(-w,w)$. The
maximum entropy power spectrum is defined as that $S(f)$ which
maximizes (3.1) subject to the constraints

$$\int_{-w}^{w} e^{-i2\pi kf} S(f)\,df = R_X(k) \tag{3.2}$$

for $-n \leq k \leq n$, n fixed. Now through the use of Lagrange
multipliers it can be shown that the requirement that (3.1) be
a maximum subject to (3.2) implies (see Burg (1975) and Chen and
Stegen (1974))

$$S(f) = \beta \left| 1 + \sum_{k=1}^{n} a_k e^{-i2\pi fk} \right|^{-2} \tag{3.3}$$

where β is a positive constant. Comparison of (3.3), the

maximum entropy power spectrum, and (2.5), the spectrum of a
discrete autoregressive process, shows they are of course the
same. However, the motivation for the two is somewhat different.
From (3.2) one can again obtain the system (2.6a,b) with
$\beta = \sigma_Z^2$. Thus one could estimate the a_i as before. This has
all the same defects previously brought out since it requires
estimation of the autocorrelations. To circumvent this problem
Burg has developed an algorithm for estimating the a_i directly.
Because in most seismic data processing, one is more interested
in estimating $Z(t)$ in (2.1) than estimating the spectrum, the
a_i are of interest in their own right. The vector $[1, a_2, \ldots,
a_n]^T$ is usually referred to as the prediction error filter (PEF)
in geophysical problems. Thus the problem of estimating the
coefficients of a discrete autoregressive process and estimating
the PEF can be considered as equivalent, and the Burg method for
estimating the PEF is of value in problems of the former formulation as well. Although the main thrust of this paper is spectral
estimation, we have mentioned this associated problem because it
is probably the most important use of the Burg method. Before
specifically describing this method we give the following brief
discussion of the Levinson algorithm.

3.1. The Levinson Algorithm

The Levinson algorithm is a recursive means for solving the
Toeplitz system (2.6a,b). Burg's form of this algorithm can be
derived as follows. We denote by $[1, a_1^{(n)}, \ldots, a_n^{(n)}]^T$ the
PEF of order n. To simplify notation we put $r_k = R_X(k)$ and
then can write the system (2.6a,b) in the form

$$\begin{bmatrix} r_0 & \cdots & r_n \\ & \cdot & \\ \cdot & \cdot & \\ \cdot & & \cdot \\ \cdot & & \\ \cdot & & \\ r_n & & r_0 \end{bmatrix} \begin{bmatrix} 1 \\ a_1^{(n)} \\ \cdot \\ \cdot \\ \cdot \\ a_n^{(n)} \end{bmatrix} = \begin{bmatrix} v_n \\ 0 \\ 0 \\ \cdot \\ \cdot \\ 0 \end{bmatrix} \qquad (3.4)$$

Because of the symmetry of the matrix in (3.4), the time reversed system is also satisfied:

$$\begin{bmatrix} r_0 & \cdots & r_n \\ & \cdot & \\ & \cdot & \\ & & \cdot \\ & & \cdot \\ r_n & \cdots & r_0 \end{bmatrix} \begin{bmatrix} a_n^{(n)} \\ \cdot \\ \cdot \\ a_1^{(n)} \\ 1 \end{bmatrix} = \begin{bmatrix} 0 \\ \cdot \\ \cdot \\ 0 \\ v_n \end{bmatrix} \tag{3.5}$$

From this it follows that

$$\begin{bmatrix} r_0 & \cdots & r_{n+1} \\ & \cdot & \\ & \cdot & \\ & & \cdot \\ & & \cdot \\ r_{n+1} & \cdots & r_0 \end{bmatrix} \left\{ \begin{bmatrix} 1 \\ a_1^{(n)} \\ \cdot \\ \cdot \\ a_n^{(n)} \\ 0 \end{bmatrix} - c_n \begin{bmatrix} 0 \\ a_n^{(n)} \\ \cdot \\ \cdot \\ a_1^{(n)} \\ 1 \end{bmatrix} \right\} = \begin{bmatrix} v_n - c_n e_n \\ 0 \\ 0 \\ \cdot \\ \cdot \\ e_n - c_n v_n \end{bmatrix} \tag{3.6}$$

where $e_n = r_{n+1} + a_1^{(n)} r_n + \ldots + a_n^{(n)} r_1$. Choosing $c_n = e_n/v_n$ we see that the right-hand side becomes $[v_{n+1}, 0, \ldots, 0]^T$ where $v_{n+1} = v_n - c_n e_n = v_n[1 - (e_n/v_n)^2]$ and so we have computed the $n + 1$ order PEF as

$$\begin{bmatrix} 1 \\ a_1^{(n+1)} \\ \cdot \\ \cdot \\ \cdot \\ a_{n+1}^{(n+1)} \end{bmatrix} = \begin{bmatrix} 1 \\ a_1^{(n)} \\ \cdot \\ \cdot \\ a_n^{(n)} \\ 0 \end{bmatrix} - c_n \begin{bmatrix} 0 \\ a_n^{(n)} \\ \cdot \\ \cdot \\ a_1^{(n)} \\ 1 \end{bmatrix} \tag{3.7}$$

The c_n are called the *partial correlation coefficients* in the statistical literature and sometimes the *reflection coefficients* by geophysicists. From the fact that $v_{n+1} \geq 0$ we conclude that $\left| c_n \right| \leq 1$. The conditions that $\left| c_k \right| < 1$, $k = 1$, ..., n, are necessary and sufficient for the nth order PEF to be minimum-phase (Burg (1975)).

The Levinson algorithm has been widely used in geophysics for calculating the PEF with the correlations r_k estimated from time series data. It has been found necessary to window the estimated correlations to obtain satisfactory PEF's (Foster (1968)), the result being that the subject is infused with art and is not very pleasing from a theoretical point of view.

3.2. Burg's Approach

Burg's goal has been to estimate the PEF directly from the data without the preliminary step of estimating the autocorrelations r_k. As many authors have done, he has proposed to choose the filter coefficients so as to minimize some measure of the prediction error. The difficulty is that the obvious ways of doing this do not usually lead to minimum-phase filters. Burg's major contribution is to see that certain features of the Levinson algorithm suggest a way out of this dilemma.

Now the actual forward prediction error at a point k for time series samples $x_1, ..., x_N$ is

$$\varepsilon_k = x_k + a_1 x_{k-1} + ... + a_n x_{k-n} \tag{3.8}$$

where $k = n + 1, n + 2, ..., N$. An obvious measure of the total error is

$$\sum_{k=n+1}^{N} \varepsilon_k^2 \tag{3.9}$$

Unfortunately, minimizing this expression with respect to the a's does not necessarily produce a minimum-phase operator, even for $n = 1$.

For stationary time series there is no preference in time direction, so it would make equally good sense to use the backward prediction error

$$\delta_k = x_k + a_1 x_{k+1} + \ldots + a_n x_{k+n} \tag{3.10}$$

$k = 1, N - n$. One would then define a measure of total prediction error as

$$\sum_{k=1}^{N-n} \delta_k^2 \tag{3.11}$$

But why not, equally well, use a sum of the two as the measure?
That is, minimize

$$\sum_{k=n+1}^{N} \varepsilon_k^2 + \sum_{k=1}^{N-n} \delta_k^2 \tag{3.12}$$

Minimization of this expression can be shown to lead to a
minimum-phase operator for $n = 1$. But for $n = 2$ and higher,
this is not necessarily the case.

It is at this point that recourse to the Levinson algorithm
is needed. We may write (3.7), which expresses the PEF of order
$(n + 1)$ in terms of the PEF of order n, as

$$a_k^{(n+1)} = a_k^{(n)} - c_n a_{n+1-k}^{(n)} \tag{3.13}$$

$k = 1, \ldots, n + 1$, where we define $a_0^{(n)} = 1$ and $a_{n+1}^{(n)} = 0$.
For $(n + 1)$ the measure (3.12) of total prediction error then
becomes

$$f(c_n) = \sum_{k=n+2}^{N} \left[x_k + \sum_{j=1}^{n+1} \left(a_j^{(n)} - c_n a_{n+1-j}^{(n)} \right) x_{k-j} \right]^2$$

$$+ \sum_{k=1}^{N-n-1} \left[x_k + \sum_{j=1}^{n+1} \left(a_j^{(n)} - c_n a_{n+1-j}^{(n)} \right) x_{k+j} \right]^2 \tag{3.14}$$

Under the assumption that the $a_j^{(n)}$ are known, Burg proposes determining c_n by minimizing this expression. The
$a_j^{(n+1)}$ are then specified by (3.13).

3.3. The Burg Algorithm

Define

$$P_k^{(n)} = x_{k+n+1} + \sum_{j=1}^{n+1} a_j^{(n)} x_{k+n+1-j} \tag{3.15a}$$

and

$$Q_k^{(n)} = x_k + \sum_{j=1}^{n+1} a_j^{(n)} x_{k+j} \tag{3.15b}$$

Then the expression (3.14) becomes

$$f(c_n) = \sum_{k=1}^{N-n-1} \left[\left(P_k^{(n)} - c_n Q_k^{(n)} \right)^2 + \left(Q_k^{(n)} - c_n P_k^{(n)} \right)^2 \right] \tag{3.16}$$

Differentiating with respect to c_n and setting the derivative to zero produces

$$c_n = \frac{2\sum_{k=1}^{N-n-1} P_k^{(n)} Q_k^{(n)}}{\sum_{k=1}^{N-n-1} \left(P_k^{(n)^2} + Q_k^{(n)^2} \right)} \tag{3.17}$$

That $\left| c_n \right| < 1$ follows from expanding the left-hand side of the inequality

$$\sum_{k=1}^{N-n-1} \left(P_k^{(n)} \pm Q_k^{(n)} \right)^2 > 0 \tag{3.18}$$

As we have seen, this guarantees the minimum-phase property. (Equality can only happen with positive probability in the deterministic case.)

Now from the defining relations (3.15a,b) for $P_k^{(n)}$ and $Q_k^{(n)}$ and the Levinson relation (3.13), we obtain the recursions

$$P_j^{(n+1)} = P_{j+1}^{(n)} - c_n Q_{j+1}^{(n)} \tag{3.19a}$$

and

$$Q_j^{(n+1)} = Q_j^{(n)} - c_n P_j^{(n)} \tag{3.19b}$$

for $j = 1, N - n - 2$.

Defining

$$S^{(n)} = \sum_{k=1}^{N-n-1} \left(P_k^{(n)^2} + Q_k^{(n)^2} \right) \tag{3.20a}$$

$$R^{(n)} = \sum_{k=1}^{N-n-1} P_k^{(n)} Q_k^{(n)} \tag{3.20b}$$

and using the relations (3.19a,b), we produce the following recursion for $S^{(n)}$.

$$S^{(n+1)} = (1 + c_n^2)S^{(n)} - 4c_n R^{(n)} - \left(Q_{N-n-1}^{(n)} - c_n P_{N-n-1}^{(n)} \right)^2$$

$$- \left(P_1^{(n)} - c_n Q_1^{(n)} \right)^2 \tag{3.21}$$

Explicitly then, the Burg algorithm is as follows. (M is the maximum length of operator desired.)

<u>Algorithm.</u> Initialization:

$$P_j^{(0)} = x_{j+1}, \quad j = 1, N - 1 \tag{3.22}$$

$$Q_j^{(0)} = x_j, \quad j = 1, N - 1 \tag{3.23}$$

and

$$S^{(0)} = x_1^2 + x_N^2 + 2 \sum_{j=2}^{N-1} x_j^2 \tag{3.24}$$

At the nth step   (n = 0, ..., M)

$$R^{(n)} = \sum_{j=1}^{N-n-1} P_j^{(n)} Q_j^{(n)} \tag{3.25}$$

and

$$c_n = 2 \frac{R^{(n)}}{S^{(n)}} \tag{3.26}$$

Recursive step from n to n + 1: $S^{(n+1)}$ from (3.21) $P^{(n+1)}$ and $Q^{(n+1)}$ from (3.19a,b).

This algorithm is highly efficient computationally and extremely simple to program. Coupled with the formula (2.7) for the estimate of the maximum entropy spectrum it is finding wide use in geophysics.

Observe that the last equation in the system (2.6a,b) can be written

$$r_n = - \sum_{j=1}^{n} a_j^{(n)} r_{n-j} \qquad (3.27)$$

If M is the maximum length of prediction error operator we have chosen to estimate, then (3.27) can be used, letting n run from 1 to M, to compute $r_1, \ldots, r_M$ provided r_0 is known. We may simply normalize by choosing $r_0 = 1$ or alternatively we may estimate r_0 as

$$r_0 = \frac{1}{N} \sum_{j=1}^{N} x_j^2 \qquad (3.28)$$

In any event, knowledge of the prediction error operators for $n \leq M$ is equivalent to knowledge of the first M autocorrelation coefficients. (Actually, as can be seen from the basic recursion (3.13), the prediction error operator of length M completely determines all the shorter operators and hence the first M correlation coefficients.)

A final comment is in order before leaving the Burg method. That is, to explore the connection with the window methods we can estimate the PEF via the Burg algorithm, the correlations from (3.27) and then compute a window from (2.8) with the estimated correlations on the left-hand side. We have done this for a limited amount of actual seismic data and find that the "Burg window" is very close to the rectangular window. (It cannot be exactly rectangular because the Burg window guarantees the minimum phase property.) Thus one would expect that the spectrum estimated via the Burg algorithm would have a minimum amount of smoothing; i.e., that the estimated spectrum would have sharp definition.

4. G-SPECTRAL ESTIMATORS

The estimate of $S(f)$ in (2.5) is of course the power spectrum of an autoregressive process of order n. There are situations for which this may not be the case.

We have seen that the maximum entropy spectral estimator and the autoregressive estimator are identical, the difference in practice being the method used to estimate the a_i. They can be expected to give their best results when the process is in fact a finite autoregressive process. This frequently may not be true. In fact, if $X(t)$ is the primary synthetic seismogram, $Z(\tau)$ is the reflectivity function, and $a(t)$ is the seismic wavelet, it can be shown (see Foster (1975)), under certain assumptions, that

$$X(t) = \int_{-\infty}^{t} a(t - \tau) Z(\tau) \, d\tau \tag{4.1}$$

where $a(t)$ is a minimum-phase kernel. But (4.1) implies $X(t)$ is a continuous autoregressive process whose order n is completely determined by $a(t)$. That is,

$$X^{(n)}(t) + a_1 X^{(n-1)}(t) + \ldots + a_n X(t) = Z(t) \tag{4.2}$$

for some set of a_i, where we assume $Z(t)$ is white and $E[X(t)Z(S)] = 0$ for $t < S$. However, if $X(t)$ is sampled, say at unit intervals, it is well known (see Jenkins and Watts (1968), Phadke and Wu (1974)) that the result is a mixed autoregressive-moving average process. More specifically,

$$X(t) + \alpha_1 X(t - 1) + \ldots + \alpha_n X(t - n)$$

$$= Z(t) + \sum_{k=1}^{n-1} \beta_k Z(t - k) \tag{4.3}$$

But the spectrum of neither (4.1) nor (4.2) is given by (2.5),
and hence neither of our previous methods seems appropriate from
the start. We now introduce a method which yields a consistent
estimate for the spectrum when (2.1), (4.1) or (4.2) is correct.
Moreover, as we shall see, this method also eliminates to a
great extent the basic problem of window methods; i.e., the need
for stable estimates of the autocorrelation for a large number
of lags. In order to facilitate our discussion in the remaining
part of the paper we now introduce the so-called G_n-transforms.

Let

$$F(t) = \int_a^t f(x) \, dx \tag{4.4}$$

Then we define the G_n-transform by

$$G_n[F(t);h] = \frac{\begin{vmatrix} F(t) & F(t+h) & \ldots & F(t+nh) \\ f(t) & f(t+h) & \ldots & f(t+nh) \\ \vdots & & & \vdots \\ f(t+(n-1)h) & & \ldots & f(t+(2n-1)h) \end{vmatrix}}{\begin{vmatrix} 1 & 1 & \ldots & 1 \\ f(t) & f(t+h) & \ldots & f(t+nh) \\ \vdots & & & \vdots \\ f(t+(n-1)h) & & \ldots & f(t+(2n-1)h) \end{vmatrix}} \tag{4.5}$$

If the denominator in (4.5) is zero at an isolated t_0, we define $G_n[F(t_0);h] = \lim\limits_{t \to t_0} [F(t);h]$ while, if (4.5) takes the

form 0/0 identically over $[a,\infty)$, we define $G_n \equiv G_{n-1}$. This

is referred to as the *drop back rule*. With regard to the G_n

transform we have the following results. [For proofs see Gray,

Atchison and McWilliams (1971).]

<u>Theorem 4.1</u>. If f satisfies the differential equation

$$y^{(n)} + \beta_1 y^{(n-1)} + \dots + \beta_n y = 0, \quad a \le x < \infty \tag{4.6}$$

for some set of β_i's, then

$$G_\gamma[F(t);h] \equiv \int_a^\infty f(x)\,dx \tag{4.7}$$

for all $t \ge a$ and $\gamma \ge n$.

If $\gamma > n$ in (4.7), it can be shown that $G_\gamma[F(t);h] \equiv 0/0$

and the drop back rule must be applied to get the equivalence in

(4.7). This fact can, as we shall discuss later, be utilized to

estimate the order of the autoregressive process or in the dis-

crete case, the length of the PEF.

In order to apply the G_n-transform it is necessary to make

another definition since G_n is a nonlinear transform. That is,

if

$$S(\omega;t) = \int_{-t}^{t} e^{-i2\pi\omega x} f(x)\,dx \tag{4.8}$$

we define

$$G_n[S(\omega;t);h] = G_n\left[\int_0^t \left(e^{-i2\pi\omega x} f(x) + e^{i2\pi\omega x} f(-x)\right) dx;h\right] \tag{4.9}$$

From Theorem 4.1 it follows that if $f(x)$ and $f(-x)$ satisfy

an equation such as (4.6) for $x > 0$, then

$$G_n[S(\omega;t);h] \equiv \int_{-\infty}^{\infty} e^{-i2\pi\omega x} f(x)\, dx = S(\omega;\infty) \qquad (4.10)$$

We demonstrate this remark by the following example.

<u>Example 4.1.</u>

Find the Fourier transform of (i) $e^{-\alpha|x|}$ and (ii) $e^{-i2\pi\alpha x}$, α real.

(i) Since $e^{-\alpha x}$ is a solution of $y' + \alpha y = 0$ for all x, we know from Theorem 4.1 (clearly $m = 2$ here) that

$$G_2[S(\omega;t);h] = S(\omega;\infty) \qquad (4.11)$$

Since we can take any $t \geq 0$ and (4.7) holds, we select $t = 0$. Then

$$G_2[S(\omega;0);h]$$

$$= 2\,\frac{\begin{vmatrix} 0 & \int_0^h e^{-\alpha x}\cos 2\Pi\omega x\, dx & \int_0^{2h} e^{-\alpha x}\cos 2\Pi\omega x\, dx \\[6pt] 1 & e^{-\alpha h}\cos 2\Pi\omega h & e^{-2\alpha h}\cos 4\Pi\omega h \\[6pt] e^{-\alpha h}\cos 2\Pi\omega h & e^{-2\alpha h}\cos 4\Pi\omega h & e^{-3\alpha h}\cos 6\Pi\omega h \end{vmatrix}}{\begin{vmatrix} 1 & 1 & 1 \\[6pt] 1 & e^{-\alpha h}\cos 2\Pi\omega h & e^{-2\alpha h}\cos 4\Pi\omega h \\[6pt] e^{-\alpha h}\cos 2\Pi\omega h & e^{-2\alpha h}\cos 4\Pi\omega h & e^{-3\alpha h}\cos 6\Pi\omega h \end{vmatrix}}$$

$$= \frac{2\alpha}{\alpha^2 + (2\Pi\omega)^2} \qquad (4.12)$$

Note that if $\omega = j/h$, $j = 0, 1, 2, \ldots,$ eq. (4.12) takes the indeterminant form $0/0$. We have only alluded to this possibility before and therefore some remarks are in order. When $G_k[F(t;\omega);h] \equiv 0/0$ (i.e., for all h), the transform is defined

by the so-called "drop back" rule; i.e., $G_k[S(\omega;t);h]$
$= G_{k-1}[S(\omega;t);h]$, $G_0[S(\omega;t);h] = S(\omega;t)$. With this convention
no difficulties arise.

(ii) Since $e^{i2\pi\alpha x}$ is a solution of $y'' + (2\pi\alpha)^2 y = 0$,
we know that

$$G_2[S(\omega;t);h] \equiv S(\omega;\infty) \tag{4.13}$$

Taking $t = 0$ we obtain

$$G_2[S(\omega;0);h] = 2\,\frac{\begin{vmatrix} 0 & \dfrac{\sin 2\Pi(\omega-\alpha)h}{2\Pi(\omega-\alpha)h} & \dfrac{\sin 4\Pi(\omega-\alpha)h}{4\Pi(\omega-\alpha)h} \\[2mm] 1 & \cos 2\Pi h(\omega-\alpha) & \cos 4\Pi h(\omega-\alpha) \\[2mm] \cos 2\Pi h(\omega-\alpha) & \cos 4\Pi h(\omega-\alpha) & \cos 6\Pi h(\omega-\alpha) \end{vmatrix}}{\begin{vmatrix} 1 & 1 & 1 \\[2mm] 1 & \cos 2\Pi h(\omega-\alpha) & \cos 4\Pi h(\omega-\alpha) \\[2mm] \cos 2\Pi h(\omega+\alpha) & \cos 4\Pi h(\omega-\alpha) & \cos 6\Pi h(\omega-\alpha) \end{vmatrix}} \tag{4.14}$$

Using the back drop rule for the case $0/0$ we obtain by (4.14)

$$G_2[S(\omega;0);h] = \begin{cases} 0 & \text{if } \omega \neq \alpha \\ \infty & \text{if } \omega = \alpha \end{cases} \tag{4.15}$$

That is,

$$G_2[S(\omega;0);h] = \delta(\omega - \alpha) \tag{4.16}$$

The above example suggests the following estimator of the spectral density. That is, let $\hat{R}_T(\tau)$ be an estimator for the autocorrelation $R(\tau)$ based on a record of length T. Then we define the G-spectral estimator as $G_k[\hat{S}(\omega);T_0]$, where

$$\hat{S}(\omega;T_0) = \int_{-T_0}^{T_0} e^{-i2\pi\omega\tau}\hat{R}_T(\tau)\,d\tau \tag{4.17}$$

T_0 fixed and $0 < T_0 < T - (2n - 1)h$.

Now note that if $X(t)$ satisfies (4.1), $R(\tau)$ satisfies

$$R^{(n)}(\tau) + a_1 R^{(n-1)}(\tau) + \ldots + a_n R(\tau) = 0 \tag{4.18}$$

and since $R(-\gamma) = R(\gamma)$ over $(-T_0, T_0)$ (assuming real processes) we have at once

$$G_{2n}[S(\omega; T_0)] = \int_{-\infty}^{\infty} e^{-i2\pi\omega\tau} R(\tau) \, d\tau \tag{4.19}$$

This result, of course, is not enjoyed by the autoregressive method, maximum entropy method or window methods. Moreover, the G_n-transform has a discrete analogue, namely the e_n-transform (see Shanks (1955)), and all of the remarks made above concerning G_n with regard to (4.1) can be made for e_n if (4.3) is the correct model. In fact, it can be shown that in the special case where the β_k in the right side of (4.3) are zero (i.e., the maximum entropy assumption holds), this discrete version of the G-spectral estimator is exactly the maximum entropy spectral estimator if the $R(\tau)$ are estimated via the Burg method, and it is the autoregressive estimator if the equations (2.8) are utilized to estimate the $R(\tau)$. Of course, if either (2.1) or the general case of (4.3) is the proper model, the G_n-spectral estimator is not equivalent to either of those methods. Moreover, in both instances of (4.1) or (4.3), the G_n satisfies the fol- lowing type consistency theorem.

<u>Theorem 4.2.</u> Suppose $\lim\limits_{T \to \infty} \hat{R}_T(\tau) = R(\tau)$ a.s. and that $R(\tau)$ satisfies (4.6) when $\tau \geq 0$ for some set of α_i . Then for any fixed T_0 and h

$$\lim\limits_{T \to \infty} G_k[\hat{S}_T(\omega; T_0); h] = S(\omega; \infty) \text{ a.s.} \tag{4.20}$$

for every $k \geq n$, n defined as in Theorem 4.1.

$\underline{\text{Proof.}}$ Under our assumptions it is clear that

$$\lim_{T\to\infty} \hat{S}_T(\omega;T_0) = \int_{-T_0}^{T_0} e^{-i2\pi\omega\tau}R(\tau)\ d\tau \quad \text{a.s.} \tag{4.21}$$

where here we have tacitly assumed the interchange of the limit with the integral over the *fixed* interval $(-T_0,T_0)$. Therefore

$$\lim_{T\to\infty} G_k[\hat{S}_T(\omega;T_0);h] = G_k[\lim_{T\to\infty} \hat{S}_T(\omega;T_0);h]$$

$$= G_k\left[\int_{-T_0}^{T_0} e^{-i2\pi\omega\tau}R(\tau)\ d\tau\right]$$

$$= \int_{-\infty}^{\infty} e^{-i2\pi\omega\tau}R(\tau)\ d\tau \tag{4.22}$$

The interchanging of the G_k with the limit follows from the fact that the denominator in (4.22) can vanish only at isolated points, in which case the numerator will also vanish or both numerator and denominator will be identically zero. The drop back rule then applies and hence we can assume the denominator in (4.22) is not zero. The discrete version of the G_n-spectral estimator can be shown to satisfy a similar theorem to the above and is currently being studied by F. W. Morgan (1975) in preparing his Ph.D. dissertation.

5. CONCLUDING REMARKS

In this paper we have attempted to describe the autoregressive and maximum entropy methods of spectral estimation. As a result, we have seen that in actuality the only difference in those methods is the manner in which one estimates the coefficients in the autoregressive process or equivalently the PEF. Of course, as we have remarked, this can be a significant difference, especially when estimation of a_i is the problem of interest rather than spectral estimation. On the other hand, when spectral estimation is the issue, we have pointed out some theoretical shortcomings of both methods and introduced the G-spectral estimator. One of the most notable characteristics of this latter estimator is that it is consistent for the spectral density independent of the length of the lag window $(-T_0, T_0)$. Thus, for sufficiently small n, one can limit himself to utilizing only estimates of $R(\tau)$ for short lags and hence eliminate a large amount of statistical variability as well as numerical computation. In addition, we remarked that this latter method can also be utilized to estimate the length of the PEF. This is true since theoretically, if $r > n$, we get $G_r[F(t);h] \equiv 0/0$. Of course, numerically this will not be exactly the case. However, if the method of Pye (1972) in the continuous case or Wynn (1956) in the discrete case is used to calculate G_r recursively, one finds that $G_n[F(t);h] \approx G_{n+1}[F(t);h] \approx \ldots$, or $G_r[F(t);h]$ begin to vary wildly for $r > n$, depending on the numerical precision involved. Thus, in either case a plot of $G_n[F(t);h]$ as a function of n will normally suffice to determine n. An in-depth study of the actual performance of G-spectral estimates, both continuous and discrete, on simulated data as compared to the autoregressive estimator, maximum entropy estimator, and window methods is currently under consideration and should appear in a future paper.

BIBLIOGRAPHY

Akaike, Hirotugu (1970). Statistical predictor identification. *Ann. Inst. Statist. Math.* **22**, 203-17.

Burg, John P. (1967). "Maximum entropy spectral analysis", presented to 37th Meeting, Society Explor. Geophys., Oklahoma City, Okla., October 31, 1967.

Burg, John P. (1975). "Maximum Entropy Spectral Analysis",
    Doctoral thesis, Stanford Univ., Stanford, California.

Chen, W. R. and Stegen, G. R. (1974). Experiments with
    maximum entropy power spectra of sinusoids. *J. Geophys.
    Res.* 79 (20).

Foster, M. R., Sengbush, R. L. and Watson, R. J. (1968).
    Use of Monte Carlo techniques in optimum design of the
    deconvolution process. *Geophysics* 33, 945.

Foster, M. R. (1975). Transmission effects in the contin-
    uous one dimensional seismic model. *Geophys. J. R.
    Astr. Soc.* (to appear).

Gray, H. L., Atchison, T. A. and McWilliams, G. V. (1971).
    Higher order G-transformations. *SIAM J. Numer. Anal.*
    8 (2).

Jenkins, G. M. and Watts, D. G. (1968). *Spectral Analysis
    and Its Applications*. San Francisco: Holden-Day, Inc.

Koopmans, L. H. (1974). *The Spectral Analysis of Time
    Series*. New York: Academic Press, Inc.

Kromer, R. E. (1969). "Asymptotic properties of the auto-
    regressive spectral estimator", Tech. Report No. 13,
    Dept. of Statistics, Stanford Univ., Stanford, Cali-
    fornia.

Morgan, F. W. (1975). "The G-Spectral Estimator", Doctoral
    thesis, Southern Methodist University, Dallas, Texas.

Parzen, E. (1969). Multiple time series modeling. *Multi-
    variate Analysis, Vol. 2* (Ed., P. R. Krishnaiah).
    New York: Academic Press, Inc.

Phadke, M. S. and Wu, S. M. (1974). Modeling of continuous
    stochastic processes from discrete observations with
    applications to sunspots data. *J. Amer. Statist. Assoc.*
    69 (346).

Pye, W. C. (1972). "Pseudoinverses and Higher Order G-trans.
    formations", Doctoral thesis, Texas Tech University,
    Lubbock, Texas.

Shanks, Daniel (1955). Nonlinear transformations of diver-
    gent and slowly convergent sequences. *J. Math. Phys.*
    34.

Shannon and Weaver (1949). *The Mathematical Theory of
    Communication*. Urbana, Ill.: Univ. of Illinois Press.

Steiglitz, K. (1964). Power spectrum identification for
    adaptive systems. *IEEE Trans. Appl. Industry* 83, 195-7.

Ulrych, T. J. (1972). Maximum entropy power spectrum of
    truncated sinusoids. *J. Geophys. Res.* 77 (8).

Wynn, P. (1956). On a device for computing the $e_m(S_n)$
    transformation. *Math. Tables Other Aids Comput.* X.

INDEX

A

Admissibility, 41-42
  admissibility conditions,
    42
    asymptotically equal, 51
    convex, 50, 52
    equal, 51
    monotone, 49, 52
    random, 52-53
    random convex, 50
    repeatable, 50, 52
    well-structured, 49, 52
Admissible discriminant
    algorithms, 41-42
Ahlberg, J. H., 33-34, 40
Akaike, H., 13, 15, 171, 187
Alphabet
  chinese, 134
  sending, 134
  receiving, 134
Anderson, T. W., 41, 47, 56,
    61, 80-81, 83, 117,
    165
A posteriori probability, 48
A priori probability, 49, 80
Aronszajn, N., 30, 31
Arseven, E., 156, 167
ARTF, 13
ARTFACT, 13
Asterisk
  innovations, 8
  model, 8
Atchison, T. A., 182, 188
Atmospheric turbulence, 18
ATR, 135
Attributes
  pattern classification by,
    121

Autocorrelation, 170, 175, 179,
    181, 184
Autoregression
  criterion transfer
    function, 13
  infinite representation, 6,
    7
  order of the, 2, 182
  scheme of order p, 4, 11
Autoregressive
  moving average, 180
  processes, 37, 169, 173, 180,
    186
  spectral estimation, 169,
    171, 180, 185, 186
  transfer functions, 13
Average transmission rate, 135
Axiomatic approach, 41

B

Backward shift operator, 5
Bahadur, R. R., 80
Barnard, M. M., 155, 165
Barnett, G. O., 56-57
Barnett, V. D., 96, 118, 157,
    165
Bartlett, M. S., 148, 150, 152,
    154-157, 165
Basic factorization, 164
Bessel functions, 147
Binet, F. E., 150, 165
Bledsoe, W. W., 127, 138, 145
Box, G. E. P., 5, 16
Browning, I., 127, 138, 145
Bryant, J., 69

192   INDEX

Burg algorithm, 171, 173,
     177-179
  method, 175, 179, 185
  window, 179
Burg, J. P., 169, 171-173,
     175-176, 187-188

C

Canonical analysis, 147, 149,
     155-156, 158-159
Canonical correlations,
     147-148, 150, 157
Carson, H., 112
Casey, R., 146
CAT, 13
Cauchy distribution, 88, 91,
     94, 96, 107
Cauchy-Schwarz inequality,
     20
Chanda, K. E., 83, 112
Chandrasekaran, B., 43, 58
Characteristic equation, 23
Chen, P., 121
Chen, W. R., 172, 187
Chernoff, H., 96, 118
CIA(m), 13
Classification procedure
  asymptotic properties,
     88
  Bayes Optimal, 61
  by attributes, 121
  efficient, 84
  maximum likelihood, 61
  nonparametric, 83, 86, 88
  optimal rule, 91
  Rule, 86
  Rule M, 85
  Rule M̂, 85, 92, 96
  Rule M*, 86, 89, 93, 96
  sampling properties, 88
  variables for, 44
Cleveland, W. S., 9, 16
Clusters, 44
Cochran, W. G., 83, 118
Collinearity, 147, 155
  factors, 148
Conditional probabilities,
     48, 53

Consistent estimator, 116
Contingency Table, 153, 158
Convex hull, 50, 52
Covariance function, 2
Covariances, 3, 159
Cover, T. M., 41, 45, 56
Criterion autoregressive
     transfer function, 13
Criterion information Akaike,
     13
Crossing probability bounds,
     17
Curve crossing, 28

D

Darcey, L., 69
Das Gupta, S., 84, 118
Davidon-Fletcher-Powell
     minimization program,
     69
Davis, H. T., 10, 16, 33, 35-36,
     40
Decision rule, 45
Decomposition, 2
Detection problems, 18
Diehr, P., 54-56
Difference operator, 37
Differential operator, 37
  linear, 33
Differentiating probability of
     misclassification, 64
Direction factors, 148
Discriminant analysis, 41-56,
     147-163
Discriminant analysis
     algorithms, 41
  average linkage, 46, 51
  Bayes-type, 41, 48, 52-55
  centroid, 46, 51
  furtherest neighbor, 46, 51
  least squares, 47, 52
  linear discriminant analysis,
     47
  nearest k-neighbor, 41, 46
  nearest neighbor, 42, 51
  quadratic discriminant
     analysis, 47
  sequential, 41, 45, 56

Distance function, 44
Doob, J. L., 37, 40
Drop back rule, 182, 184, 186
Dummy variable, 149, 151, 153

E

Earthquakes, 19-20, 28
Egyptian skulls, 155
Eigenvalue, 156
Eigen-vector, 156, 158
Eisenbeis, R. A., 152, 166
Energy of the process, 24
e_n-transform, 185
Entropy, 127
Estimation of parameters, 5

F

Factorial experiments, 156
Failure of buildings, 19
Fatigue mechanism, 29
Feature selection, 61, 127
Feinstein, 134
Fisher, L., 41, 45-47, 50,
        53-58
Fisher, R. A., 69, 80, 148,
        154
Fix, E., 84, 118
Fletcher-Powell minimization
        program, 69
Fortier, J. J., 47, 57
Foster, M. R., 169, 175, 180,
        188
Fourier transform, 183
Friedman, H. P., 47, 57
Fubini theorem, 30
Fu, K. S., 44, 57
Fukunaga, K., 69, 76-79

G

Garry, G. A., 56-57
Gastwirth, J. L., 96, 118
Gateaux differential, 63
Geometric approach to
        discriminant analysis,
        43
Gilbert, E. S., 55, 57, 145

Gilbert, G. G., 152, 166
G_n-spectral estimator, 185-186
G_n-transform, 181-182, 185
Goodness of fit, 148
Govindarajulu, Z., 84, 118
Gower, J. C., 41, 57
Grams, W. F., 118
Gray, H. L., 169, 182, 188
Green, P., 156, 166
Greens function, 37
Grenander, U., 10, 16, 43, 57
G-spectral estimator, 169, 180,
        184-185, 187
Gupta, A. K., 84, 118
Gupta, S. S., 118
Guseman, L. F., Jr., 61, 63-65,
        80-81

H

Halbert, M., 156, 166
Handwritten chinese, 121
Hart, P. E., 41, 45, 56
Hilbert-Schmidt kernel, 24
Hills, M., 53, 57
Hodges, J., 84, 118
Hoeffding, W., 115, 118
Homer, F., 81
Hopkins, C. D., 83, 118
Hotelling's generalized T^2, 147
Howell, L. J., 18, 31
h(t), 17
Hudimoto, H., 84, 88, 119
Hudimoto's rule, 84
Hypothesis of independence,
        162-163

I

Impulse response function, 17,
        29
Innovation, 2
Input process, 17

J

Jardine, N., 41, 57
Jenkins, G. M., 5, 16, 180, 188
Johns, M. V., 96, 118

Johnson, S. C., 41, 49, 58
Jones, R. H., 10, 16
Jung, J., 96, 119

K

Kanal, L., 43, 58
Kimeldorf, G., 34, 40
Koopmans, L. H., 17, 19-20,
        27-28, 31, 171, 188
Kromer, R. E., 12, 15-16, 171,
        188
Kronmal, R., 54-56
Kshirsagar, A. M., 147, 149,
        150, 155-159, 166,
        167
Kuiper, K., 45-46, 58

L

Lachenbruch, P. A., 53, 58,
        83, 119
Lag window, 171, 187
Ledley, R., 55, 58
Lee, J. C., 83
Levinson algorithm, 172-173,
        175-177
Lewis, T., 157, 165
Lincoln, T., 55, 58
Linear filter, 21
Linear system, 17
Linguistic model for
        discriminant
        analysis, 43-44
Lin, Y. K., 18, 31
Loève, M., 26, 31
Logistic, 88, 91, 94
Loss function, 47
Luenberger, D. G., 63, 66, 81
Lusted, L., 55, 58

M

Markov chain, 156, 158
Markov renewal processes, 156
Maximum entropy, 185-186
    estimation, 169
    method, 185
    power spectrum, 172-173

spectral estimator, 180, 185
Maximum likelihood classifica-
        tion, 61
McWilliams, G. V., 182, 188
Meijer's G-function, 147
Mellin-Barnes type integrals,
        147
Mickey, M. R., 53, 58
Minimum mean square error
        linear predictability,
        2
Minimum mean square prediction
        error, 3
    linear, 2
Minimum phase property, 170,
        175-177, 180
Misclassification
    derivative of probability of,
        64
    probability of, 53, 61
Misspecification, 7
Mixed autoregressive-moving
        average scheme, 4-6
Modeling, 1
    hard, 1
    identification, 5
    misspecification, 7
    soft, 1
Monte Carlo simulation, 21
Morgan, F. W., 186, 188
Mosaic, 127, 134
Moving average scheme of order
        q, 4
    infinite, 170
    infinite representation, 6-7
Multivariate analysis, 159
Multivariate methods, 147

N

Nagy, G., 146
NASA/Johnson space center, 69
Nilson, E. N., 33-34, 40
Nonparametric
    classification rules, 83
    density estimation, 48
    test procedures, 83
Nonstationary Gaussian process,
        17

Normal distribution, 88, 91,
      94, 159
Null functions, 154

P

Parker, R., 55, 58
Partial correlation
      coefficients, 174
Parzen, E., 1, 16, 29-31, 38,
      40, 48, 58, 169,
      171, 188
Patel, H. I., 156, 158, 167
Patrick, E. A., 55, 58
Pattern classification, 43-44
      121, 148
PEF, 170, 173-176, 179,
      186-187
Peridontal conditions, 156
Phadke, M. S., 180, 189
Pickands, J., III, 19, 31
Pillai's criterion, 147
Power spectrum, 170, 172, 180
Prediction, 1, 5, 11
   best linear, 38-39
   coefficients, 3
   finite memory, 3
   m-memory, 4
   multivariate, 33
Prediction error, 176
   backward, 176
   filter, 170, 173, 176
   forward, 175
   operator, 179
Priestley, M. B., 23, 31
Primary synthetic seismogram,
      180
Primitives, 44
Probability
   a posteriori, 48
   a priori, 49, 80
   bounds, 17, 24, 28
   crossing, 17, 29
Probability of misclassifica-
      tion, 53, 61
   derivative of, 64
Pye, W. C., 187, 189

Q

Qualls, C., 17, 20, 27-28, 31

R

Rao, C. R., 83-84, 119, 155,
      167
Recognition scheme, 138
Reflection coefficients, 174
Reflectivity function, 180
Regression, 1, 149
Reproducing kernel, 30
Revo, L. T., 83, 119
Riesz, F., 24, 31
Rosenblatt, M., 10, 16
Roy's largest root, 147
Rubin, H., 165
Rubin, J., 41, 47, 57-58

S

Sandia image digitizer, 121
Schatzoff, M., 160, 167
Seismic
   data processing, 173
   wavelet, 188
Selection of variables, 151
Sengbush, R. L., 188
Serfling, R. J., 118
Shanks, D., 185, 189
Shannon, 172, 189
Sibson, R., 41, 57
Sneeringer, C., 83, 119
Solomon, H., 47, 57
Sorum, M., 53, 58, 83, 86, 119
Specht, D. F., 52, 56, 58
Spectral density function, 2, 6
   estimator, 169, 173, 184
Spectral decomposition, 21
Spectrum, 173, 179, 181
   maximum entropy, 179
Spline, 33
   cubic, 35, 40
   draftsman's, 35
   interpolatory, 39
   L-34-36, 38-39

Srikantan, K. S., 167
Stationary process, 21
Stationary time series, 11
Statistical model for
        discriminant
        analysis, 43-44
Steck, G. P., 121, 138, 145
Stegen, G. R., 172, 188
Steiglitz, K., 169, 171, 189
Stochastic process, 169
Stoller, D. C., 84, 119
Structural approach to
        discriminant
        analysis, 43
SUBROUTINE DFMFP, 69
Supervised pattern recognition,
        44
Sz.-Nagy, B., 24, 31

                T

Tapering function, 21
Taxa, 44
Taylor's data, 154
Test of hypothesis on roots,
        160
Time series, 1
    stationary, 2
Toeplitz system, 173
Training data, 45
Transfer functions, 5, 6, 21
Transmission rate, 134, 139

                U

Uhr, L., 138, 145
Ulrych, T. J., 172, 189

                V

Van Ness, J., 41, 47, 50, 53,
        57, 59

Van Ryzin, J., 48, 59
Variables for classification,
        44
Variance-covariance matrix, 159

                W

Wahba, G., 34, 40
Waknis, M. N., 118
Walker, H. F., 61, 63-65, 80-81
Walsh, J. L., 33-34, 40
Watson, G. S., 150, 165
Watson, R. J., 188
Watts, D. G., 180, 188
Weaver, 172, 189
Whitening filter, 170
White noise, 4, 170, 180
    process, 169
Wijsman, R., 24
Wilcoxon test, 88
Wilks' Λ, 147-148, 154-155,
        157-159
    factorization, 147
Williams, E. J., 148-150, 152,
        155-157, 167-168
Wilson, W. J., 158
Window, 175, 179
    rectangular, 171
    spectral estimation, 179,
        181, 185
Wishart distribution, 159
Wright, W. E., 41, 59
Wu, S. M., 180, 189
Wynn, P., 187, 189

                Y

Yao, J. T. P., 20-21, 27-28, 31
Yule, G. V., 150
Yule-Walker equations, 3

                Z

Zonal polynomials, 147